BOGS OF THE NORTHEAST

Charles W. Johnson

With the Advice and Assistance of Ian A. Worley

Illustrated by Meredith Edgcomb Young

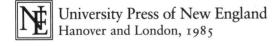

University Press of New England
Hanover and London, 1985

University Press of New England
Brandeis University
Brown University
Clark University
University of Connecticut
Dartmouth College
University of New Hampshire
University of Rhode Island
Tufts University
University of Vermont

Copyright 1985 by Charles W. Johnson

Printed in the United States of America

Library of Congress Cataloging in Publication Data

Johnson, Charles W., 1943–
   Bogs of the Northeast.
   Bibliography: p.
   Includes index.
   1. Bog ecology—Northeastern States.   2. Bogs—
Northeastern States.   1. Worley, Ian A.   11. Title.
QH104.5.N58J64   1985   574.5'26325'0974   84–40587
ISBN 0–87451–325–1
ISBN 0–87451–331–6 (pbk.)

Source for poem on page 4: Excerpt from "Bogland" from *Poems 1965–1975* by Seamus Heaney. Copyright © 1969, 1980 by Seamus Heaney. Reprinted by permission of Farrar, Straus and Giroux, Inc.

We acknowledge the generous financial assistance of the Alden Trust in the production of this book.

CONTENTS

PREFACE

I can't remember exactly what I saw the first time I went to a bog, but I do recall my wonder at the strangeness of it all. Suddenly I had entered a new world. Perhaps my sensations were primed by what I had read about bogs, but most came as original revelations. And most of what impressed me *was* sensation: the spongy undulations of the mat, the subsidence of each of my steps, the subtle yet arresting colors of rain-glazed sphagnum mosses, the gnarled forms of seemingly ageless plants. Nor was my imagination cheated: The day was noiseless and draped with storm clouds; a fog will-o'-the-wisped up and over a leaden pond; the place seemed to throb with deep energy, buried under a gray mantle of solitude. Some strange gods had to be living there.

That first encounter plucked some sonorous chord within me, and subsequent trips to other bogs, whatever their size or the season, never dulled the magic of that first song. From then on bogs were my unearthly music, and I listened. In the years since, I have studied bogs both scientifically and casually, trying to understand the maker of the music. In this search for understanding I have been like anyone else who cherishes a piece of the world as his own, wanting still to experience the special feelings yet seeking to know more.

No matter how much we know or think we know about bogs, we cannot fully appreciate them if we only read the score and never really listen to the music. Mountains, oceans, forests, and bogs are more than places to those who love them. They have hearts like hibernating animals, beating imperceptibly yet vitally. Some people can sense a pulse. Accordingly, I look on this book as a biography of the bogs in one section of North America. Bogs keep diaries from their birth, recorded page upon page in their peats. Their histories remain secrets for hundreds or thousands of years, until we literally dig deeply to reveal them. This book, though, goes beyond strict scientific measures. It is a natural history of the "personalities" and relationships produced by the plants and animals that help make bogs living systems.

The fact that men have walked on the moon does not mean that poets can no longer write about it; it is still a sacred part of the heavens. So with bogs on earth. While the scientist in us seeks to understand them, the poet in us wants to keep them away from complete discovery, safe in some shadow of mystery.

ACKNOWLEDGMENTS

There are so many people to thank for their help in the making of a book. Some gave you the idea in the first place. Some lent support when you needed it; without them you would have given up in dismay long ago. Some allowed you the time to work on it, sacrificing their own solitude. Some supplied information. Some reviewed it and must have been frustrated to tears or laughter. Some did all of these things. Ian Worley was one who did all. As teacher, colleague, and friend, he has been with me all along, from the first burst of interest through the drudgery of revisions to the publication of this book. He too "feels" the bogs and listens to what they have to say, and he shared his perceptions with me. He was also always the quiet and concerned counselor. In a way, this is as much his book as mine.

The days spent afield with good friends have been the most powerful stimulus for this book. With friends there is no need to explain what it means to see an orchid blooming new in the muted shade of a cedar tree, the wind winnowing through stalks of sedges, or the tamaracks feathered yellow in the fall. They know what it means. For all those times and discoveries, for the given gifts of sharing, I want to thank especially Dick Andrus, bryologist and energetic ecologist, whose idea of fun is to stand knee-deep in muck, staring at sphagnum mosses, in the rain; Marc DesMeules, herpetologist, naturalist, and conservationist, who always seems to know what stone to turn over to find salamanders, while I flip all the others; Bob Spear, naturalist and bird carver, who can find a bird in any place or in any piece of wood; and Stephan Syz, wetland lover and philosopher, who doesn't mind getting up at 4 : 00 A.M. to get to a bog, to photograph an orchid as the first light of day hits its petals.

Other people in the Northeast have helped by showing me sites; by providing information, photographs, or a place to stay; or just by being companions and friends: Leslie Allen, Tim Barnet, Ross Bell, Charlie Cogbill, Bill Countryman, Annie Crowe, Richard Czaplinski, Gordon Day, Mike DiNunzio, Nona Estrin, John Feingold, Ed Flaccus, Keith Gentzler, Bill Gove, Dave Hartman, April Hensel, Peter Hudson, Bob Jervis, Bob Klein, Peter Ludwig, Les Mehrhoff, Nora Mitchell, Nancy Murray, Ann Pesiri, Chuck Racine, Tom Rawinski, Tudor Richards, Jane Schlossberg, Tom Siccama, Henry Stopka, Liz Thompson, Celeste Tracey, Hank Tyler, Hub Vogelmann, Margaret Watkins, Anne Winchester, Hank Webster, Henry Woolsey, Robert Zampella, and Peter Zika.

I am grateful to many organizations that made their people and resources available to me. I particularly want to thank the various state conservation agencies, universities, the National Park Service, the Maine Critical Areas Program, and The Nature Conservancy for their assistance, and the University of Vermont and the Vermont Agricultural Experiment Station for the loan of cameras and other equipment.

Dick Andrus, Anton Damman, and Margaret Johnson (who is also my mother), reviewed earlier versions of the manuscript. I deeply appreciate their encouragement and criticisms, and needed both. Meredith Edgcomb Young has been a sensitive illustrator. Donna Pollard made sense out of snipped-and-taped paragraphs, voluminous corrections, and inky smudges, then skillfully and patiently typed a fine copy. (I hope she still likes bogs.)

Above all, the members of my family, near and far, are with me always. My sons, Hunter, Brendan, and Graylyn, are still my sunrise, bringing light to every darkness.

East Montpelier, Vermont   C. W. J.
February 1985

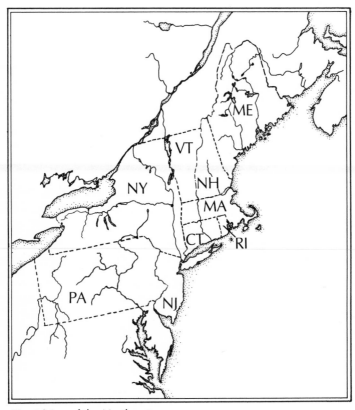

Fig. 1 Map of the Northeast.

BOGS OF THE NORTHEAST

Fig. 2 *Moose Bog, Vermont. (Photo by Charles W. Johnson.)*

1 | PERCEPTIONS OF BOGS

Over great spans of time, space, and cultures, humans have viewed bogs in many different ways. Some have considered them from a purely practical standpoint, focusing on what can be taken from them, how to use them, or how to get around or eliminate them in the interests of "progress." Others have been wary and often apprehensive of them. Still others have been eager and curious, seeking unusual plants. Yet even those trying to be purely objective have usually felt some emotion, some wonder.

Like other parts of our earth, bogs hold—and yield—things people want. Where they are a dominant part of the landscape, bogs tend to be intimately entwined with the lives of people. For millennia the Danes and the Irish (among others) have hand-cut blocks of peat from bogs to heat their homes and cook their meals. For centuries the Finns have harvested bogs for cloudberries, which they make into a delicate liqueur. More recently the Russians have mined peat in enormous quantities to feed giant electrical generating plants. In the New World bogs provided Native Americans with many plants and plant parts for food, medicines, magic charms, tools, trinkets, and even toys for children. In what was to become Massachusetts early European settlers quickly learned to prize the bog-grown cranberry and to turn other wetlands into artificial bogs for the cultivation of this crop. Later, midwestern and Ontario bogs were drained and converted into other croplands. Further south bogs and streams in the New Jersey Pine Barrens were mined for iron as early as the 1770s. Today Canadians and Americans scoop great quantities of peat from bogs for use in gardens, potting soil, and the little pots in which we start spring vegetables.

To people who have gone to them in search of food, fuel, and other products, bogs have not been loathsome or fearsome, as they have often been depicted in stories. Instead, they have been part of an inheritance, familiar from birth on. To them, bogs have been much the same as a hardwood forest is to one who cuts firewood from it, or a lake to one who catches dinners there, or a pine barrens to one who goes to it for blueberries in the summer. Usefulness has been intrinsic to value.

But there has been another dimension to the human relationship with bogs. The sentient dimension. The almost spiritual. The most celebrated indication of a spiritual connection between humans and bogs came with the discovery of bodies of Bronze and Iron Age people (from 1500 to 500 B.C.) preserved in northern European bogs. These bodies, some of them extraordinarily well kept—except for the bones, which were dissolved in the acidic waters—are the subject of a fascinating book, *The Bog People*, in which the Dane Peter Glob unfolds the discoveries, discusses the sometimes grisly details of how the victims died, and hypothesizes about why the bodies were buried in the bogs.

Although he acknowledges that anthropologists disagree about the beliefs and rituals of the bog people, Glob theorizes that these early Germanic cultures regarded at least some bogs as places sacred to the gods, particularly to the goddess of fertility, Nerthus (whom we might modernize by calling "Mother Earth"). Her favor could make the difference between life and death to these inhabitants of marginal agricultural lands of the North, where crop failure could mean disaster for the entire coming year. Bogs were the outdoor "temples" where humans, perhaps criminals condemned to death anyway, were sacrificed to appease Nerthus and in return were given the honor of a form of immortality:

They were sacrificed and placed in the sacred bogs; and consummated by their death the rites which ensured for the peasant community luck and fertility in the coming year. At the same time, through their sacrificial deaths, they were themselves consecrated for all time. (Glob 1969)

In the sense that these peoples hoped to benefit from the rituals they performed, the bogs used for these burials were "producing" something important. But we must wonder why bogs were considered sacred places. Did they appear to be spots where the wrath of the gods and goddesses boiled to the surface? Did they inspire reverence or demand homage? Or did these primitive people sense something divine in places where dead things might not decay and thus where, in a manner of speaking, the dead lived on? Whatever the reasons, the bog people certainly had powerful religious sentiments tied directly to bogs.

Later cultures continued to react to bogs with strong feelings of disquietude, awe, or reverence, expressed in their literature. Thus Shakespeare in *The Tempest* (1611):

All the infections that the sun sucks up
From bogs, fens, flats, on Prosper fall and make him
By inchmeal a disease!

Carl Linnaeus in his diary (1732):

Shortly afterwards began the muskegs [bogs], which mostly stood under water; these we had to cross for miles; think with what misery, every step up to our knees. The whole of this land of the Lapps was mostly muskeg. . . . Never can the priest so describe hell, because it is no worse. Never have poets been able to picture Styx so foul, since that is no fouler.

And Sir Arthur Conan Doyle in *The Hound of the Basker-villes* (1902):

Rank reeds and lush, slimy water-plants sent an odour of decay and a heavy miasmatic vapour onto our faces, while a false step plunged us more than once thigh-deep into the dark, quivering mire, which shook for yards in soft undulations around our feet. Its tenacious grip plucked at our heels as we walked, and when we sank into it it was as if some malignant hand was tugging us down into those obscene depths, so grim and purposeful was the clutch in which it held us.

For poets and novelists, bogs have been sources of somber inspiration. Thus Emily Dickinson around 1861:

> How dreary—to be—Somebody!
> How public—like a Frog—
> To tell one's name—the livelong June—
> To an admiring Bog!

And Seamus Heaney in "Bogland" (1969):

> . . . Our unfenced country
> Is bog that keeps crusting
> Between the sights of the sun.
> . . . .
> Only the waterlogged trunks
> Of great firs, soft as pulp.
> Our pioneers keep striking
> Inwards and downwards,
>
> Every layer they strip
> Seems camped on before.
> The bogholes might be Atlantic seepage.
> The wet centre is bottomless.

Even today, with all our accumulated knowledge and scientific attitudes to dispel myths, we remain somewhat ambivalent about bogs. We find them intriguing, yet we shun them as somewhat peculiar. They remain mysterious—neither solid land nor water but a realm in between. The plants are unusual, the terrain generally uninviting. An aura of spirits still emanates from them to stir our imaginations . . . how desolate when the winds sweep across these moors, through stunted and spindly spruce trees, while the fog rolls over like heavy dark drapes.

We are still impressed, and duly apprehensive, when we hear stories of cows or horses disappearing into bogs or of great gaping holes waiting to engulf the unwary traveler. We persistently recall this sogginess in everyday language as we become "bogged down" in this or that. Perhaps the feelings are innate: one young boy visiting a bog for the first time described it with distaste as "a water bed with garbage floating on top." One of my sons, when I carried him as a two-year-old over a particularly bouncy quaking bog, gripped me tightly and in a beseeching voice cried, "Daddy, I no like these bumps!"

The bogs of a continent, a region, or even a smaller locale differ greatly in physical form, vegetation, and utility to people. The "turf" cut by the Irish and Scottish for home heating may be firm and relatively dry. A quaking bog in New England may be sodden, unstable, and used only for observation or occasional meanders by the curious. We have realized that bogs are incredible bits of the world, valuable both for what is living at the surface and for what is stored below in the peats. We are beginning to appreciate them simply for what they are: discrete ecosystems unlike any others.

Botanists have long sought out bogs as havens for rare plants, islands of significant biota. But in the last half-century, with the perfection of scientific techniques permitting the reconstruction of ancient environments through the analysis of deposited peat, diverse researchers have discovered treasure chests of information packed and preserved in the peats— from ecologists, paleoecologists, and climatologists to anthropologists and even historians. In fact, more than any other ecosystem, bogs are invaluable to our understanding of past climates, vegetation, wildlife, and even human life. Revelations from bog studies, fascinating to laymen and scientists alike, have provided some of the most important advances in our knowledge of changes in our world's landscapes. Currently ecologists in many nations, perceiving bogs as complete natural systems, explore their subtle dynamics, establishing new, rich ecological understandings and rejecting long-accepted yet erroneous notions.

Scientific knowledge, however real and crystalline, has not displaced entirely the spiritual associations bogs held with prehistoric cultures. For we retain a kind of reverence, just as potent as primitive worship—sensitivity toward their special biological identity, their intricate functions, their variety, and their delicacy yet age-old permanence.

2 | BASIC TERMS AND DEFINITIONS

Bog is an old and well-worn word, probably dating back to medieval times and derived from the Celtic *bocc*, meaning "soft." It is applied to a wide variety of soggy, moist, spongy, or otherwise wet areas over a great many regions of the world. Recently, however, with the extensive development of peat resources in many countries and with worldwide advances in ecology, the term has acquired multiple meanings. Often it refers to but one specific type of wetland landscape.

Still the preferred lay term in the Northeast, in scientific language *bog* has been replaced by the more descriptive if less romantic *peatland*. Simply, peatlands are lands whose soils are peat—the partially or incompletely decomposed remains of dead plants and, to some extent, animals. However, peatlands in the ecological sense are much more than lands or soils alone. Ian Worley has provided one good, comprehensive definition:

Peatlands are three-dimensional portions of the earth's landscape which are wetlands and have organic soils; they include the full depth of organic materials, regardless of origin; they include all waters within or on top of the organic materials; and they include all

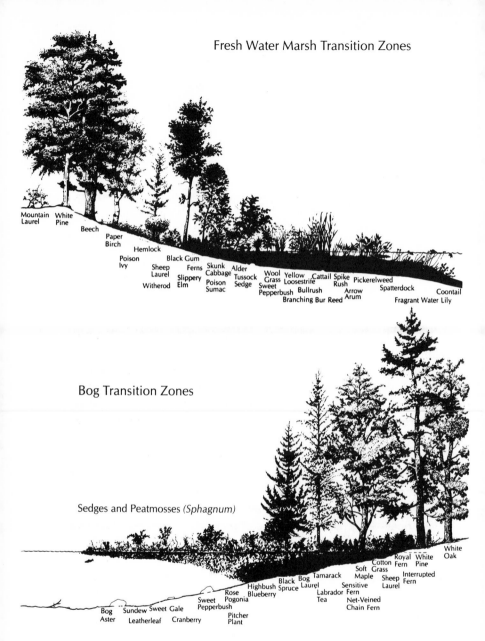

Fresh Water Marsh Transition Zones

Mountain Laurel
White Pine
Beech
Paper Birch
Hemlock
Poison Ivy
Sheep Laurel
Witherod
Slippery Elm
Black Gum
Ferns
Skunk Cabbage
Poison Sumac
Alder
Tussock Sedge
Sweet Pepperbush
Wool Grass
Branching Bur Reed
Yellow Loosestrife
Bullrush
Cattail
Arum
Spike Rush
Arrow
Pickerelweed
Spatterdock
Fragrant Water Lily
Coontail

Bog Transition Zones

Sedges and Peatmosses *(Sphagnum)*

Bog Aster
Sundew
Leatherleaf
Sweet Gale
Cranberry
Sweet Pepperbush
Rose Pogonia
Pitcher Plant
Highbush Blueberry
Black Spruce
Bog Laurel
Labrador Tea
Tamarack
Sensitive Fern
Net-Veined Chain Fern
Soft Maple
Cotton Grass
Royal Fern
Sheep Laurel
White Pine
Interrupted Fern
White Oak

Figs. 3 and 4 *Marsh and bog.* Marsh and bog vegetation (generalized) bordering ponds and lakes in southern New England. Many of the species shown occur throughout the Northeast. Marsh soils typically contain much clay or inorganic silts, whereas nearly pure organic deposits accumulate beneath many bog mats. *(Reprinted with permission from Environmental Protection Agency, 1981, Region 1, New England Wetlands: Plant Identification and Protective Laws. The plants are not drawn to scale.)*

organisms living within or atop the organic materials and water. (Worley 1981)

This definition acknowledges that peatlands are identifiable functioning ecosystems containing soils, waters, and communities of living things.

As it is used in Worley's definition (which has been derived from that of the U.S. Fish and Wildlife Service), *wetlands* has many possible meanings: lands where the water table is at, near, or above the surface long enough during the growing season to promote the formation of special (hydric) soils or to support the growth of hydrophytes (special water-loving plants).

Peatlands, then, are only one type of wetland. Other types exist where peat is less abundant or is absent. For instance, marshes are wetlands with herbaceous emergent plants, and swamps are wetlands with perennial woody vegetation—trees, shrubs, or both. Marshes and swamps have predominantly mineral (nonorganic) soil, even though much organic matter may be incorporated. (*Marsh* and *swamp* are terms with various local meanings. I use them here as they are most commonly employed in the Northeast.)

The U.S. Soil Conservation Service defines *organic soil* in a precise though rather complex way. A greatly simplified definition is soil that has at least 12 percent organic carbon (though most have much more, in some cases up to 98 percent); the remaining components can be clay, sand, or other minerals. Often a field investigator can identify organic soil by feel; in a lab the dried organic soil burns.

This basic definition of peatland can be applied throughout the world, despite differences in the climates and vegetation that have formed the peat. For convenience and diagnosis, peatlands in the Northeast are today often classified according to two different systems. One considers how and where the peatland develops; the other is based on the water and nutrient status of the peatland. The terms of both systems are largely interchangeable.

Fig. 5 *Marsh*. Emergent and floating-leaved herbaceous plants predominate in a marsh. Shown here are spatterdock (*Nuphar variegatum*) in the foreground, common cattail (*Typha latifolia*) at the right, and common arrowhead (*Sagittaria latifolia*) beyond. *(Photo by Ian Worley.)*

Fig. 6 *Swamp.* Silver maple (*Acer saccharinum*) grow in a swamp along Lake Champlain. Swamp white oak (*Quercus bicolor*) is another common tree of such swamps. Note the high-water mark (dried clay and silt) on the tree trunks. *(Photo by Charles W. Johnson.)*

Fig. 7 *Kettlehole bog*. A typical kettlehole bog, more or less classic in structure, in a small New England basin. This bog has the often-cited concentric zonation of a central pond, a floating mat of peat mosses, sedges, and dwarf shrubs, bordered by taller shrubs and subdued trees, and an enclosing bog forest of conifers. *(Photo by Charles W. Johnson.)*

A peatland that is confined to a basin of some kind and does not extend beyond or above the physical limits of that basin is called a *level* or *flat* peatland (although in fact there may be no such thing as a perfectly level peatland). The basin holds the water in which the peatland develops; usually there is little water inflow or outflow. The classic "quaking bogs" of the glaciated regions are examples of level peatlands: they are characteristically small and often sit in a kettlehole.

Many peatlands do not form discrete basins but instead occupy shallow depressions on gentle gradients, usually in association with some kind of water discharge, such as springs, slow-moving brooks, and seepages. There are many kinds of these *slope* peatlands, which vary in their plant and animal life, hydrology, and size. Formed throughout the Northeast, in the extreme north they may be large (up to several hundreds of acres) with their surfaces remarkably patterned by alternating ridges and hollows.

In parts of the Northeast (and elsewhere in the world) the peatlands actually are raised above their surroundings; their peats have accumulated to such an extent that the peatland surface extends well above and sometimes well beyond the boundary of the original basin and water table. *Raised* peatlands, as far as we know, are restricted in the Northeast to northern and coastal Maine and the northern Adirondacks of New York. These are the southernmost raised peatlands in a zone extending northward and eastward into Canada. Among the largest of our peatlands (ranging from several hundred acres to, in one case, more than 4,000 acres), their distinctive shapes and surface patterns depend on their proximity to the ocean.

A second way of classifying peatlands recognizes ecologically their important water and nutrient characteristics. *Geogenous* (earth-originating) peatlands are those receiving water principally from underground and/or surface sources, waters that have picked up nutrients from their passage through soils and bedrock. These are also called *minerotrophic* (mineral-nourished) peatlands. (Another, older term is "eutrophic.") Because moving water also flushes away acids and other metabolic by-products, geogenous peatlands are the least acidic. Most slope peatlands are geogenous, as are marshes, swamps, and tidal estuaries.

Ombrogenous (rain-originating) peatlands, on the other hand, receive their water from precipitation—rain, fog, or snow. These are one and the same as raised bogs, whose surfaces are above the influences of ground and surface waters. Ombrogenous peatlands, also called *ombrotrophic* (rain-fed),

Fig. 8 *Raised bog*. A medium-size raised bog in south-central Maine. Central portions of this unpatterned bog rise several feet above the soggier margins; thus, nutrients are supplied only from the air. In such nutrient-poor environments the infrequent trees are much stunted, and the hummocky bog surface is dominated by peat mosses and dwarf shrubs. *(Photo courtesy of Maine Critical Areas Program.)*

are the poorest in nutrients of all peatlands, for they must depend on what is brought in by precipitation and air rather than on water that has been "fertilized" in the ground. With little nutrient input and flushing by water, these are the most acid of peatlands.

Finally, some peatlands are called *transitional* or *oligotrophic* (poorly fed), being between geogenous and ombrogenous in their richness in nutrients. Many peatlands belong to this rather "catchall" category, and ecologists have created

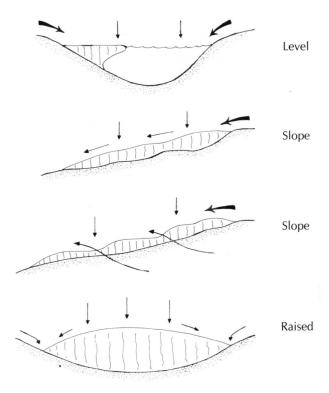

Level

Slope

Slope

Raised

Fig. 9 *Nutrient sources.* The arrows indicate the main sources of water and nutrients in the different types of peatlands. *(Adapted from Worley, 1981.)*

numerous subdivisions based on nutrient availability, acidity, and vegetation.

Now back to the word *bog.* Scientists now generally restrict the term *bog* to ombrotrophic and very oligotrophic peatlands and call minerotrophic and less oligotrophic ones *fens.* In the strictest usages, *bog* is reserved for only raised peatlands (i.e., peatlands whose nutrient source is only the atmosphere); similarly, *fen* is reserved for strongly enriched peatlands. Intermediates between these two extremes bear many names and generally are less well studied. In this book *peatland* is the preferred and most frequently used term. *Bog* is often synonymous with peatland; but except when the context is quite clear, *bog* refers specifically to just the nutrient-poor raised (ombrotrophic) peatlands.

The Northeast contains a great spectrum of peatland types, from ombrotrophic (raised) bogs and oligotrophic (level) bogs to oligotrophic (poor or extremely poor) fens and minerotrophic (rich) fens. Many different types frequently exist within one large peatland area (these are then called peatland *complexes*) or in conjunction with other wetlands.

There is also a special nomenclature for particular parts of peatlands. At its perimeter a peatland often has a *moat* (or *lagg*, after the Swedish), a very wet zone where water from the adjacent upland collects and flows slowly around the main peat mass. Often strikingly different from either the surrounding land or the peatland further inward, this zone can be a morass of shrubs and murky waters, an effective deterrent to many a would-be peatland visitor and thus a protector of its solitude.

Because the interlacing roots and underground stems (rhizomes) of peatland vegetation form a walkable, often somewhat springy surface, it is known as the *mat*. Whether floating on top of open water or, more typically, resting directly on peat, this mat is the interface between living plants and organic deposition. The term *quaking bog* is derived from the "walking on waves" feeling one encounters while traversing floating mats at pond margins.

At a larger scale the raised and slope peatlands of northern boreal regions exhibit special features often best seen using high-elevation photographs. The elevated portions of raised bogs have two principal regions: a sloping shoulder (*rand*) from the lagg inward and the main central mass, which may be a convex *dome* or flattened plateau. Some raised peatlands have small pondlets or wet depressions on the peat surface, either scattered seemingly at random over the surface or in

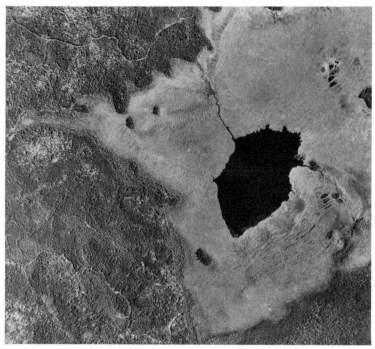

Fig. 10 *Patterned peatlands*. High-elevation photography reveals pronounced "ribbing" in 1,700-acre Number 5 Bog in western Maine. Within areas of ribbing the darker elongated features are either pools or wet depressions. The central pond is nearly 100 acres in size.

discernible concentric or eccentric patterns. Pathways, sometimes tens of feet wide, of shallow water flow on the peatland surface or between abutting domes—easily seen on aerial photographs—and are recognizable on the ground by their wetness and different species. *Soaks* emanate from the elevated portions of a raised peatland, and *water tracks* typically carry enriched water through an otherwise nutrient-poor bog.

Peatlands of boreal and arctic regions characteristically have prominent surface patterning when viewed from the air. In boreal Maine and New York peatlands, alternations of ridges and wet depressions (or pools) occur concentrically on raised bogs and parallel on sloping fens. The ridges are called *ribs* or *strings*, the pools *flarks*.

To the north of our region, in southeastern Newfoundland, *blanket* peatlands, similar to the blanket bogs of Ireland and Scotland, follow the undulations of the land, up hills and down valleys. Farther north and to our west, peatlands differ in form and vegetation, especially in the subarctic and arctic, where permafrost strongly affects peatland character. But the special terminology of these peatlands, as well as those of the subtropics and tropics, are beyond the scope of this book.

Local terms may differ from those given here. From southern New Hampshire along coastal Maine, *heath* (pronounced "hayth" in down east Maine) is used for large, usually raised, treeless bogs. *Spong* is a New Jersey Pine Barrens expression for bog, and *cripple* is their local term for a cedar (Atlantic white cedar) peatland or swamp. South of our region along the Atlantic, *pocosin* is the name for the common bog type; to the north in much of Canada and Alaska, *muskeg* is the almost universal word for peatlands, this term of Indian origin referring to wet areas of mosses, dwarf shrubs, and (usually) some scattered conifers.

With these stepping stones of terms and definitions, let us now continue our journey of exploration into the fascinating world of peatlands.

3 | THE GENESIS OF PEATLANDS

Most of us in the United States have been taught that bogs are rare things—small, fragile, hidden in deep and cold pockets of the terrain, and restricted to the northern regions of the continent. In some parts of the world, however, peatlands are neither rare nor small: they may occur as continuous tracts for hundreds, even thousands, of square miles! It has been estimated that peatlands worldwide amount to nearly two million square miles, with Canada having the most (626,200 square miles), the USSR second (579,000 square miles), and the United States third (including Alaska, 115,800 square miles) (see Table 1).

Peatlands are by no means limited to cold or glaciated regions, for they exist at all but the highest latitudes—in as seemingly unlikely locations as Cuba, New Zealand, Paraguay, Indonesia, Pakistan, and Israel. In Jamaica there are peatlands below sea level, in East Africa on high mountains, in South America among rain forests.

Wherever peatlands occur, within broad themes they are unusually alike in structure and development, sometimes differing markedly only in the constituent plant species. What do all of these areas have in common that allows peatlands to develop with such similarities when they represent such dif-

TABLE I *Estimate of Total Peatlands, by Country*

| Country | Peatland Area | | |
	Millions of Acres	Millions of Hectares	Percentage of World Total
Canada	425	170	40.36
USSR	375	150	35.62
USA (including Alaska)	75	30	7.12
Indonesia	65	26	6.17
Finland	26	10.4	2.47
USA (not including Alaska)	25.6	10.24	2.43
Sweden	17.5	7.0	1.66
China	8.70	3.48	0.83
Norway	7.50	3.0	0.71
Malaysia	5.90	2.36	0.56
Great Britain	3.95	1.58	0.38
Poland	3.38	1.35	0.32
Ireland	2.95	1.18	0.28
West Germany	2.78	1.11	0.26
Iceland	2.50	1.0	0.24
East Germany	1.38	0.550	0.18
Cuba	1.13	0.450	0.11
Netherlands	0.63	0.250	—
Japan	0.50	0.200	—
New Zealand	0.38	0.150	—
Denmark	0.30	0.120	—
Italy	0.30	0.120	—
Hungary	0.25	0.100	—
Yugoslavia	0.25	0.100	—
Uruguay	0.25	0.100	—
France	0.23	0.090	—
Switzerland	0.14	0.055	—
Argentina	0.11	0.045	—
Czechoslovakia	0.08	0.031	—
Austria	0.06	0.022	—
Belgium	0.05	0.018	—
Australia (Queensland)	0.04	0.015	—
Romania	0.02	0.007	—
Spain	0.02	0.006	—
Israel	0.01	0.005	—
Greece	0.01	0.005	—
Bulgaria	—	0.001	—

SOURCE: Kivinen and Pakarinen (1981).
NOTE: The peatlands of most developing countries and of subtropical and tropical regions are poorly known and thus underrepresented in this table.

ferent places, climates, and vegetation in the world? What is the shared common denominator?

In theory, peatlands can develop anywhere there is a net gain in organic matter over time. Decomposition (decay), however slow or rapid, cannot exceed production if peatlands are to form and develop. The one overriding condition that slows decomposition is water—there must be wetness for peat to accumulate! Water, whether from atmospheric or groundwater sources, must be in surplus over the course of the yearly cycle. In water, especially in quiet water, dead plants and animals decompose at a fraction of the rate they do when exposed to both air and moisture—witness the preservation of old wooden ships long ago sunk in lakes and seas. If oxygen is totally lacking, organic materials cannot decompose completely. Depending on the degree of decomposition, different kinds of peat become the residue.

Not only must there be water, but it must in some way be impeded in its movement through the site. Slowed water flow allows two things to happen: first, the water, being less agitated, eventually becomes oxygen poor as dead matter decomposes within it; second, organic matter that would be borne away in currents accumulates, building the peat deposit.

A combination of climate and topography permits suitable hydrologic conditions. Basins where water collects and where outflow is restricted are good candidates for peatland formation. So may be nearly flat surfaces, but only where the regional groundwater is elevated *or* humidity is very high *or* precipitation is great throughout the growing season. Thus, even mountains and other sloping surfaces can support peatlands if there are no prolonged periods of drought or drying. The source of water, the nutrients it carries, the timing of wetting and drying, and the overall temperature regime strongly influence the kind of peatland development—bog, fen, marsh, swamp, or some combination of these.

One ecological myth that has persisted over the years is that peatlands act like huge sponges, holding onto vast quantities of water and releasing it slowly to drainage streams dur-

Fig. 11 *Powers of preservation*. The author holds a cedar post that has been in a bog for at least 30 years. The continuously exposed top part is highly weathered; the lower portion, which has been continuously submerged, is in virtually the same condition as when it was first sunk. The middle section, where water fluctuation has been greatest, shows the most rotting. *(Photo by Charles W. Johnson.)*

ing dry periods. This concept is attractive because it ascribes certain beneficial values to peatlands. Research has shown, however, that this theory is only partly accurate. Peatlands contain a great deal of water, but they do not act as thirst quenchers for water-starved areas nearby. In fact, in most cases water movement is exceedingly slow below the uppermost few inches to a foot or so of a peatland. Water can freely circulate only through the uppermost layer, the so-called active zone (or *acrotelm*). It is bound tenaciously in the intra- and intercellular spaces of the lower peats (the *catotelm*), especially when these peats are more decomposed. For some peatlands, it has been estimated that less than one percent of all the water that leaves does so below the active zone. Consequently, when the water table drops during dry periods, the peatland does not give up its waters to other areas. Instead, the active zone becomes desiccated, and the waters of the lower peats are jealously retained.

The peat-forming process is less dependent on the species of plants growing on the surface. The main contributor to peat in Maine's raised bogs is sphagnum moss; in oligotrophic basin bogs of southern New England it may be the leaves, twigs, and roots of dwarf shrubs; whereas in the Florida Everglades (a great deal of which is peatland) it is sawgrass, a sedge of the genus *Cladium*. Overall, it is the physical setting, not the biological one, that determines where peatlands can develop.

Given a satisfactory physical setting, peatlands can develop by two different but related processes: *lakefill* and *paludification*.

Lakefill simply means that a lake or pond is gradually displaced by vegetation in a sequential and generally predictable manner. For example, a typical New England kettlehole bog begins in a glacially made depression in sands and gravels. Water will collect in the depression if the groundwater table is higher than the basin bottom or if sedimentation (clay or perhaps even organics) seals the bottom. With time, the remains of algae and shoreline life settle to the bottom, beginning the gradual process of lakefill.

Sedges (family Cyperaceae) or other emergent shrubs or herbs depending on water quality and geography may colonize the perimeter of the water body, while submergent and floating-leaved plants invade the more open waters. In some kettles, as the slow buildup of organic sediments makes the water ever more shallow, shoreline growth, rooting in the near-surface peats, can extend from the edge toward the center of the pond. In other ponds floating mats of vegetation expand over open water while the slow but persistent organic

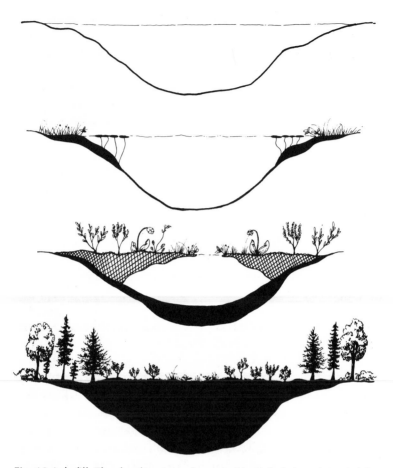

Fig. 12 *Lakefill*. The development of a typical kettlehole bog through lake-fill, much simplified and generalized. *Top*: The initial water body; plants restricted to the bordering upland. *Second from top*: With time, sediments (black) from algae and shoreline plants accumulate; floating-leaved and emergent plants border the shore as the water becomes shallower. *Second from bottom*: With increased encroachment and the associated changes in water chemistry and temperatures, a floating mat of sedges, mosses, or shrubs may form, adding more sediments and a somewhat less wet growing surface. *Bottom*: As it grows to maturity, the mat will cover the entire water surface, becoming thicker and more compacted. Bottom peats pile up from the rain-down of organic matter from above until eventually the pond is completely displaced. If nutrients and aeration are acceptable, even trees may grow on the peatland.

accumulation continues beneath. As pond becomes wetland, other communities may form where nutrients are most limited, except heaths (family Ericaceae) and peat mosses (genus *Sphagnum*); where surface waters provide enrichment, communities may be thick with sedges, herbs, and taller shrubs. Buoyant, the mat of plants and debris floats within the water body much as a saturated sponge sits in a sink of dishwater.

As bottom sediments fill the basin and as the mat matures in thickness, stability, and expanse, it becomes suitable (water levels and chemistry permitting) for taller trees. Red maple (*Acer rubrum*) and northern white cedar (*Thuja occidentalis*) can grow at richer sites, tamarack (*Larix laricina*) and black spruce (*Picea mariana*) at poorer and more northerly locations, and Atlantic white cedar (*Chamaecyparis thyoides*) near the Atlantic coast. Ultimately, according to the antiquated theory of Hydrarch Succession, the kettle completely fills and the area becomes dry land. The area fills but never ceases to be wetland; for even where deposition extends above the original height of pond water, it is with peat, and it brings the water with it! The entire process may be very slow; some kettlehole peatlands are still in early stages of development after more than ten thousand years.

Paludification, a word coined in Europe in the 1930s, refers to the creation of a peatland by the drowning or submerging of upland. Local paludification may result from altered drainage (from beaver damming, for example) or the lateral expansion of raised peatlands (as in Maine). Regional paludification, the widespread elevation of water tables and expansion of wetlands, is usually caused by changes in regional or global macroclimate (as happened in North America about 3,000 years ago, with a shift from a warmer, drier episode to a cooler, wetter one).

In our area, uplands typically are wooded. With the rising waters from either local or regional causes, the inundated roots of upland plants suffocate from oxygen starvation. Peatland paludification is nowhere more graphic than where one finds treeless raised bog transgressing on forest, the tall dead trees testimony to the upland past, the miniature bog plants the conquering invaders.

Paludification is somewhat at odds with a lingering concept within the theory of Hydrarch Succession. This concept states that all peatlands are a stage in the transition of open water to forested upland. But through paludification peatlands replace already-established upland communities! The evidence is strong that paludification is a prime mechanism for peatland development. Indeed, it appears that only a small percentage of the world's peatlands have had their beginnings as a lake or pond. Moreover, research in many parts of the world has shown that peatlands do not follow a single predictable sequence from open water to sedges and grasses to oligotrophic bog to forest. The developmental pathways vary with geographical location, water quality, climate and climatic change, topography, and now with man's influence as well. Each peatland has its own individuality within the broad themes; there may be both lakefill and paludification, or perhaps some combination of both; there may be "final" peatland stages either wooded or treeless—most, it would appear, are combinations, variable in space and time.

So as the theory of lakes to peatlands to upland is laid to rest, and as we realize the extent of paludification worldwide, should we evoke a theorem of widespread paludification and peatland climax? The well-known wetland ecologist Miron Heinselman, in his classic studies on the great Lake Agassiz Peatlands of Minnesota concluded: "One is tempted to claim a 'muskeg climax' asserting a general trend toward paludification, the very opposite of the Clementsian view." But this he tempers by following it immediately with: "But even this trend cannot persist indefinitely because climate and physiography set limits on peat accumulation" (Heinselman 1970).

Indeed, all peatlands, however small, however immense, are ultimately restricted by topographic, climatic, hydrologic, or geologic factors.

4 | FENS AND BOGS

As peatlands develop and change, water continues to be a powerful controller, certainly in its abundance but especially in its quality. Factors such as nutrient content, acidity, and oxygen richness fine-tune the peatland type and determine what kind of peatland will arise and what kind of plants will grow there. It is these water-related factors that separate the two fundamental peatland types—*bog* and *fen*. Bogs are atmospherically fed and thus poor in minerals and water flow; they are quite acidic and notably poor in species; in our area the peat mosses (*Sphagnum* species) are the principal peat formers. Fens receive surface and groundwater, and they vary greatly with the kind and degree of water enrichment. Ranging from somewhat acidic to basic, they may be dominated by sedges (family Cyperaceae), shrubs, or trees, and are typically rich in species. Plants are excellent indicators of the chemical and physical conditions in a peatland.

FENS

As water journeys down from higher ground, it either passes over and through the soil as surface flow or moves deeper into the subsoil and bedrock to become groundwater. In their

TABLE 2 *General Bog and Fen Characteristics*

	Bog	Fen
Water and Nutrient Source	precipitation	precipitation, surface water, groundwater
Available Minerals	low	moderate to high
Acidity	acidic (pH < 4.2)	slightly acidic to basic (pH > 4.8)
Principal Peat Source	sphagnum	sedges, woody plants
Species Diversity	low	low to high

SOURCE: Adapted from Minnesota Department of Natural Resources (1981).

passage these waters gather and carry organic materials, dissolved minerals and gases, and other substances.

The geomorphology of an area (structure and the "lay of the land") largely determines what substances are available for the water to carry and how fast the water travels. Both factors have a substantial bearing on a peatland's identity. Consider a young peatland only moderately to poorly developed. If the reservoir is open (as opposed to a closed basin such as a glacial kettlehole), with at least a modest outlet, the water works its way slowly through the mass, then exits. In doing so it offers its nutrients and oxygen to the plants growing on the surface. It also flushes away many of the byproducts of the plants' metabolism and decay, including acids and tannins, which inhibit plant growth.

The geology of a region has considerable influence on the quality of water. If, for example, the water passes through a limestone belt in Connecticut or Vermont, it may pick up a lot of calcium, since limestone is fairly soluble in water. This water is "sweet" or "hard" and thus tends to be basic (rather than acidic). If the water is moving slowly, it has more time to dissolve calcium than if it were coursing down a steep hill-

TABLE 3 *Some Fen and Bog Indicators of the Northeast*

	Fen (minerotrophic)	Bog (oligotrophic/ombrotrophic)
Trees	Tamarack (*Larix laricina*) Northern white cedar (*Thuja occidentalis*) Red maple (*Acer rubrum*)	Black spruce (*Picea mariana*) Atlantic white cedar (*Chamaecyparis thyoides*)
Shrubs	Bog rosemary (*Andromeda glaucophylla*) Black chokeberry (*Aronia melanocarpa*) Highbush blueberry (*Vaccinium corymbosum*) Bog birch (*Betula pumila*) Large cranberry (*Vaccinium macrocarpon*) Shrubby cinquefoil (*Potentilla fruticosa*) Sweet gale (*Myrica gale*) Swamp azalea (*Rhododendron viscosum*)	Bog laurel (*Kalmia polifolia*) Labrador tea (*Ledum groenlandicum*) Leatherleaf (*Chamaedaphne calyculata*) Small cranberry (*Vaccinium oxycoccos*) Dwarf huckleberry (*Gaylussacia dumosa*) Black crowberry (*Empetrum nigrum*) Sweet pepperbush (*Clethra alnifolia*)
Herbaceous Plants	Some sedges (e.g., *Carex oligosperma, Cladium mariscoides*) Some sphagnum mosses (e.g., *Sphagnum magellanicum*) Bladderworts (*Utricularia* spp.) Spatulate-leaved sundew (*Drosera intermedia*) Buckbean (*Menyanthes trifoliata*) Three-leaved Solomon's seal (*Smilacina trifolia*) Brook lobelia (*Lobelia kalmii*) Marsh cinquefoil (*Potentilla palustris*) Water avens (*Geum rivale*) Blue flag (*Iris versicolor*) Skunk cabbage (*Symplocarpus foetidus*)	Some sedges (e.g., *Eriophorum spissum, Carex pauciflora*) Some sphagnum mosses (e.g., *Sphagnum fuscum*) Round-leaved sundew (*Drosera rotundifolia*) Cloudberry (*Rubus chamaemorus*) Lichens (*Cladonia* spp.)

NOTE: Plants are not listed by region, nor are all species strictly confined to specific habitats in all instances.

Fig. 13 *Fen.* A Vermont fen has loose sphagnum moss hummocks for islands amid the sedges, shrubs, and watery openings, typical of many northeastern fens. *(Photo by Charles W. Johnson.)*

side. On the other hand, if the water is moving through Pennsylvania sandstone, it picks up fewer minerals. Through the granitic bedrock of the New York Adirondacks or the White Mountains of Maine and New Hampshire, it has even less opportunity to be enriched with minerals and thus stays more acidic, since granite dissolves very little in water. Often the kind of peatland that develops depends on a fine balance between the rate of flow and nutrient load of its waters. For example, the amount of dissolved oxygen or calcium available to a specific plant depends on its concentration in the water and the amount of water flowing per unit time.

Most fens in the Northeast occur where water is at the surface, discharging, or moving on downhill. Although the slopes may be imperceptibly slight, there is water movement.

Fig. 14 *Fen in winter.* Snow usually protects all low vegetation during mid-winter cold at this northern Vermont fen. The scattered, tall tamaracks (*Larix laricina*) are subjected to desiccating and frigid winds, as well as occasional mid- or late-winter ice storms. *(Photo by Richard Czaplinski.)*

In some places stream flow is evident. Fens may colonize the depressions around, below, or even on top of the discharge; numerous fens develop in small ravines, along rivers or streams in shallow valleys, and in open-end basins where the groundwater is impounded before escaping through a pinched-off outlet.

Northeastern fens most often have unpatterned zones of vegetation related to nutrients and water movement; they are wooded or open, wet, and usually lush with many kinds of plants. In the northern, noncoastal half of Maine and the adjacent Canadian provinces are string or ribbed fens—large, minerotrophic peatlands (or portions thereof) in which vegetation and surface pools of water are arranged in striking patterns. The ribs may be as much as twenty to thirty feet wide and one to three feet high. In cross section, they may look like broad staircases, with the ribs themselves as the risers and the water (flarks) the flat steps.

In North America and in Scandinavia (where ribbed fens are called *aapamires*), climate seems to be largely responsible for their distribution. The process of formation is far from clear and may involve the interaction of several forces. Two causes, however, seem to play major roles: the physical movement of water within or through the peatland, and the growth (and decay) of the vegetation in response to that movement. The alignment of the ribs is at right angles to the slope, reminiscent of the way sand dunes form perpendicular to the wind, or sand riffles to water currents. The vegetation—sphagnum mosses, shrubs, sedges, even trees—and the peat itself are in bands at right angles to the direction of water movement. Unlike sand dunes or riffles, however, the peat and vegetation are not pushed downhill by water. Rather, they are stationary and grow (and die) in those patterns.

The plant communities that constitute fens vary greatly, depending on water movement, nutrients, biogeography, climate, and local geology and topography. Generally, in the Northeast, rich (highly minerotrophic or geogenous) fens have few or no sphagnum mosses. Sedges (family Cyperaceae) dominate. Of many species, they can form impressive and attractive grasslike "pastures" and are the major peat formers. In some rich fens, shrubs and trees may be prolific, the species depending on the geographic location, especially latitude (Table 4).

Rich fens, though scarce in noncalcareous sections of the Northeast, contain an extraordinary diversity of plants, many of which are unusual. Here, otherwise rare calcium-loving wetland plants arise in sometimes astonishing profusion. Indeed, many species are on various state lists of rare and endangered plants. (It is this quality that makes rich fens so vulnerable to exploitation and overvisitation—and the reason I have identified none by name or location in this book.) Many

TABLE 4 *Some Southern and Northern Peatland Plants in the Northeast*

Southern	Northern
(Restricted to or generally more common in southern New England, the unglaciated Northeast, and/or the Atlantic Coastal Plain)	(Restricted to or generally more common in glaciated New England and New York)
Curly grass fern (*Schizaea pusilla*)	Black spruce (*Picea mariana*)
Netted chain fern (*Woodwardia areolata*)	Tamarack (*Larix laricina*)
Virginia chain fern (*Woodwardia virginica*)	Northern white cedar (*Thuja occidentalis*)
Atlantic white cedar (*Chamaecyparis thyoides*)	Pod-grass (*Scheuchzeria palustris*)
Wool-grass (*Scirpus longii*)	Hare's-tail cotton grass (*Eriophorum spissum*)
Arrow arum (*Peltandra virginica*)	Mud sedge (*Carex limosa*)
Pipewort (*Eriocaulon decangulare*)	Stunted bog sedge (*Carex paupercula*)
Yellow-eyed grass (*Xyris caroliniana*)	Few-flowered sedge (*Carex pauciflora*)
False asphodel (*Tofieldia racemosa*)	Sheathed sedge (*Carex vaginata*)
Golden-crest (*Lophiola americana*)	Bog asphodel (*Tofieldia glutinosa*)
Crested fringed orchid (*Platanthera cristata*)	White bog orchis (*Platanthera dilatata*)
Wax myrtle (*Myrica cerifera*)	Autumn willow (*Salix serissima*)
Sweet bay magnolia (*Magnolia virginiana*)	Bog willow (*Salix pedicularis*)
Poison sumac (*Toxicodendron vernix*)	Bog birch (*Betula pumila*)
Eryngo (*Eryngium aquaticum*)	Linear-leaved sundew (*Drosera linearis*)
Sweet pepperbush (*Clethra alnifolia*)	Labrador tea (*Ledum groenlandicum*)
Rose bay rhododendron (*Rhododendron maximum*)	Velvetleaf blueberry (*Vaccinium myrtilloides*)
Swamp azalea (*Rhododendron viscosum*)	Robbin's ragwort (*Senecio robbinsii*)
Maleberry (*Lyonia ligustrina*)	

NOTE: Some of these species grow in nonpeatland habitats as well.

Fig. 15 *Bog clubmoss (Lycopodium inundatum).* This small nonflowering plant occurs in acidic peatlands and uplands. Its erect, spore-bearing portions are only a few inches high. It may be found in boreal regions, on exposed sandbanks or roadsides or on the bare peats of bogs. The running stems (rhizomes) can spread over a considerable distance.

of these appealing plants are orchids, such as showy lady's slipper (*Cypripedium reginae*), rose pogonia (*Pogonia ophioglossoides*), and bog twayblade (*Liparis loeselii*), but many are of other groups—for example, grass-of-Parnassus (*Parnassia glauca*), false asphodel (*Tofieldia glutinosa*), and brook lobelia (*Lobelia kalmii*), among scores of others.

Fens of lesser nutrient input (poor fens) are probably the most common peatland type throughout the Northeast. Of many expressions, they occur in both glaciated and nonglaciated landscapes. They also can occur within or as part of swamps, and they can grade into marshes. They can be predominantly open (with herbaceous plants), shrubby, or even forested. Like rich fens, their floras often are markedly different from south to north and from inland to maritimes. Many sedges, for example, are not found at all in the north. Also, Atlantic white cedar, also called southern white cedar (*Chamaecyparis thyoides*), is relatively abundant in swamps and peatlands in the southern coastal plain, becoming less so farther north. In southeastern Maine and New Hampshire, there are only a handful of sites containing Atlantic white cedar.

BOGS

As an actively growing peatland surface becomes more isolated from surface or groundwater, the water-borne nutrients can no longer reach the plants. Deprivation of calcium, potassium, phosphorus, and other substances (most of which come from the dissolution and weathering of rocks) limits the growth of plants. By association, the flora also changes dramatically, as do all ecosystem functions. Nutrient-demanding species become progressively stunted and meager and are eventually supplanted by species tolerant of this more stringent environment. The peatland becomes less minerotrophic, more oligotrophic.

As the waters bring less, airborne life-supporting substances become increasingly more important to the bog: dust, soil particles, minerals (such as magnesium and sodium), organic matter in the form of pollen and spores. Bogs near the sea also get extra doses of salts in spray, particularly potassium, sodium, and chlorine. Air pollutants—the nitrous and sulfuric oxides and heavy metals—also make their way through the air and are deposited in peatlands, with potentially important ramifications.

Occurring in tandem with oxygen depletion in bogs is the accumulation of acids, the by-products of normal respiratory metabolism. Those formed by the breakdown of plant tissue are called humic acids and may consist of a variety of organic compounds. These give a dark color to bog waters.

The acids further suppress the metabolic action of the decomposers. In terrestrial environments, fens, and marshes, acids typically do not build to such influential proportions because they are flushed away by the waters passing through. But in bogs, where the waters cannot perform this cleansing function, the acids contribute to the slowing down of decay processes.

An observer can see, feel, and even smell the difference in decomposition rates and sterility between a bog and a more nutrient-rich wetland, such as a marsh. Besides the obvious differences in plant species, in a marsh organic matter breaks down quickly into indistinct particles; the bog retains chunks of plants or, in some cases, entire components such as leaves or seeds. When a marsh is stirred up, it emits an odor associated with decay—the gases of relatively quick-rotting substances. Because of arrested decomposition, a bog has little or no odor.

Level Bogs

The most common type of bog in the Northeast, and the one most studied in this region, is the level bog—acidic, oligotrophic, usually small to medium in size (ranging from less than an acre to around 200 acres), and occupying a basin formed in association with some glacial activity. Level bogs are found throughout the glaciated Northeast in almost any setting: small depressions at high elevations (for example, atop Mount Mansfield, Vermont, at 4,000 feet and near the summit of Mount Washington, New Hampshire, at around 6,000 feet), old pond sites at intermediate elevations, and lowland kettles in sandy-gravelly outwash plains.

Many bogs that originated as ponds are deep; some are deeper than forty feet. Many have the overall appearance of the celebrated quaking or kettlehole bogs—small, roughly circular, cradled in the embrace of a forest, with bouncy sphagnum mats either spanning the entire area or surrounding the small remnant of the once larger pond. There are, however, different expressions in shape, depth, size, and consistency.

At least two major groups of plants are typical of level bogs: sphagnum mosses and heaths. Sphagnum spreads the carpet; it is the preponderant ground cover and becomes the main component of the bog peats when it dies. The heaths are

Fig. 16 *Bog.* The entrapped waters of a kettle have allowed formation of a typical northern New England bog, with sphagnum mosses and heaths (here mostly leatherleaf, *Chamaedaphne calyculata*) predominating on the mat encircling the central pond. *(Photo by Marc DesMeules.)*

well represented in most bogs, and their robustness (or lack of it) is a good indicator of nutrient status. Tall, large-leaved, thrifty plants indicate relative mineral enrichment; short, small-leaved, dwarfed plants point to relative mineral poverty.

The vegetation sometimes occurs in clear zonal patterns, from the edge inward. Nearest the upland border, shrubs and trees may grow densely, with black spruce (*Picea mariana*), tamarack (*Larix laricina*), mountain holly (*Nemopanthus mucronata*), leatherleaf (*Chamaedaphne calyculata*), occasionally red maple (*Acer rubrum*), and other "swampy" hardwoods. Farther out on the mat the heaths and sphagna may predominate, sharing the space with a few open-growing plants such as sundews (*Drosera* species), pitcher plants (*Sar-*

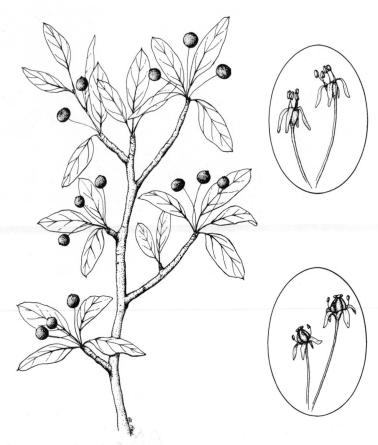

Fig. 17 *Mountain holly* (*Nemopanthus mucronata*). Mountain holly, related to (but little resembling) the familiar American holly (*Ilex opaca*),
grows in swamps, forested peatlands, shoresides, and other damp places
throughout the Northeast. The rather unusual flowers have either greatly
reduced female parts (pistils) with fertile stamens (*top inset*) or sterile
stamens and fertile pistils (*bottom inset*). Thus, there are effectively separate male and female flowers. The showy fruit has a most beautiful soft
crimson color.

racenia purpurea), and a sedge or two. In a bog with a pond, the area where the mat reaches out into the water may be colonized with lush growths of heaths and sedges. This zonation has traditionally been interpreted as an indication of Hydrarch Succession (Chapter 3), but in fact it may be more a reflection of nutrient and water conditions in particular sections of the bog.

Many other species representing many different families grow in level bogs. As a rule the species found in bogs are less diverse, less rare, and more "northern" than in fens. Level bogs, however, have much local variation in species, mostly as a result of geologic or hydrologic influences. There is also a noticeable difference in the species present in the northern sections of the glaciated Northeast from those in the southern (oak and pine forest) region. This difference is presumably tied to climate.

Raised Bogs

Climate seems to play a big role in the formation of bogs. Under certain conditions, a bog continues to develop upward, piling peat on top of peat, until it creates a mound above (and often beyond) the original basin. If this happens, the bog has passed from geogenous, through transitional, into ombrogenous. It has become the most extreme expression of a bog in the Northeast: a raised bog.

How raised bogs actually grow is a topic of considerable debate. One theory is that hummocks build upward and coalesce, "squeezing out" the hollows. The new hummocks thus occupy the sites of old hollows, but in doing so they create new hollows between themselves. Harry Godwin (1981) describes the process with a vivid analogy: "the bog surface can be thought of as porridge on a slow boil, fresh bubbles [hummocks] constantly arising in one place after another and then subsiding." Other theories assume that hummocks and hollows build up indefinitely without replacing one another; or that ridges and depressions remain in place for great

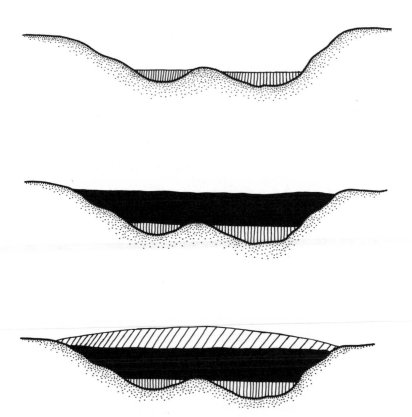

Fig. 18 *Level and raised bogs*. *Top*: The simplest level peatlands occupy or fill basins to a level controlled by existing groundwater tables. *Middle*: In some climates level peatlands can locally raise water tables to the limits of the physical basin; this may consolidate two or more smaller basins beneath a single peatland surface. *Bottom*: In even more humid and precipitation-rich climates, where precipitation nearly equals or exceeds losses by evaporation and plant transpiration, raised bogs may form over existing peatland and even on adjacent upland—above former water tables and above basin limits. Here the peatland surface receives nutrients and water from the atmosphere only!

lengths of time, even thousands of years, with neither regeneration nor replacement occurring; or that some hollows actually "consume" hummocks or some hummocks override hollows.

Of course, the growth of a raised bog must slow and stop at some point; otherwise it would become a mountain or engulf its surrounding territory. As the bog rises, so does the water level in the peat, becoming higher in the middle than at the edges, like the top of the mat. But at some stage the sphagnum surface seems to outgrow its ability to hold water against gravity, and development stops at this point. Even extremely moist atmospheric conditions cannot make up for the vertical separation from water below.

Raised bogs can form only in very wet climatic conditions. In the Northeast, only the extreme northern and northern coastal sections appear to provide these conditions. The oceanic climate of downeast coastal Maine and the Canadian maritime provinces is markedly different from that just inland. There is a greater moisture surplus owing to higher humidity during the growing season, more cloud and fog cover, less snow, and cooler summer temperatures. These conditions combine to make more water available throughout the summer (no long dry periods), principally by reducing water loss through soil evaporation and plant evapotranspiration. In contrast, peatlands in continental climates often go through periods of summer drought that counteract a potentially more vigorous sphagnum growth over a longer growing season. They also tend to receive more snow and to keep it longer than the coastal peatlands.

A comparable ecological change takes place as one moves from temperate to subarctic regions, from south to north in the northern hemisphere. But in this instance, coldness substitutes for moisture by sharply reducing the amount of time plants are exposed to possible drying conditions. Coldness simply shortens the growing season, which is usually wet anyway.

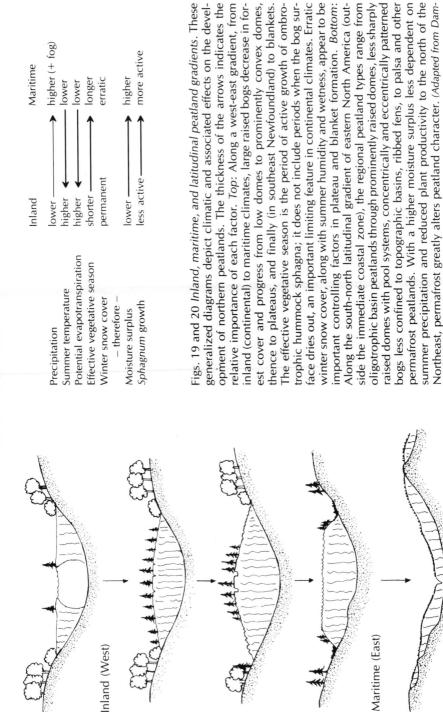

	Inland		Maritime
Precipitation	lower	→	higher (+ fog)
Summer temperature	higher	←	lower
Potential evapotranspiration	higher	←	lower
Effective vegetative season	shorter	→	longer
Winter snow cover	permanent		erratic
— therefore —			
Moisture surplus	lower	→	higher
Sphagnum growth	less active	→	more active

Inland (West)

Maritime (East)

Figs. 19 and 20 *Inland, maritime, and latitudinal peatland gradients*. These generalized diagrams depict climatic and associated effects on the development of northern peatlands. The thickness of the arrows indicates the relative importance of each factor. *Top*: Along a west-east gradient, from inland (continental) to maritime climates, large raised bogs decrease in forest cover and progress from low domes to prominently convex domes, thence to plateaus, and finally (in southeast Newfoundland) to blankets. The effective vegetative season is the period of active growth of ombrotrophic hummock sphagna; it does not include periods when the bog surface dries out, an important limiting feature in continental climates. Erratic winter snow cover, along with summer humidity and wetness, appear to be important controlling factors in plateau and blanket formation. *Bottom*: Along the south-north latitudinal gradient of eastern North America (outside the immediate coastal zone), the regional peatland types range from oligotrophic basin peatlands through prominently raised domes, less sharply raised domes with pool systems, concentrically and eccentrically patterned bogs less confined to topographic basins, ribbed fens, to palsa and other permafrost peatlands. With a higher moisture surplus less dependent on summer precipitation and reduced plant productivity to the north of the Northeast, permafrost greatly alters peatland character. (*Adapted from Damman 1979.*)

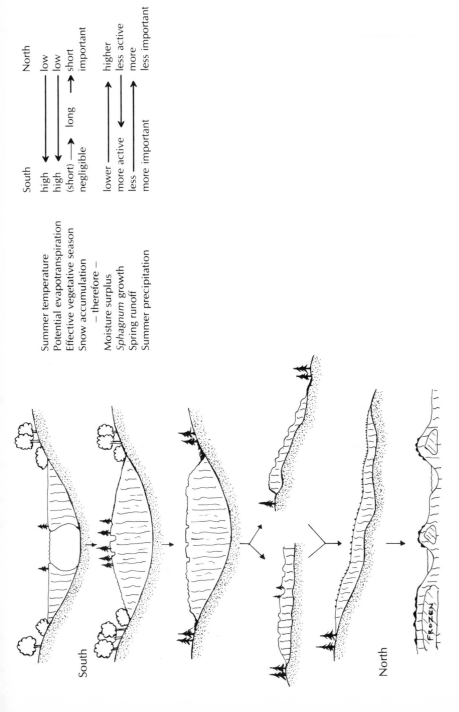

	South			North
Summer temperature	high		→	low
Potential evapotranspiration	high		→	low
Effective vegetative season	(short)	→ long →	short	
Snow accumulation	negligible		→	important
— therefore —				
Moisture surplus	lower		→	higher
Sphagnum growth	more active		→	less active
Spring runoff	less		→	more
Summer precipitation	more important		→	less important

South

North

FROZEN

TABLE 5 *Some Differences Between Inland and Maritime Raised Bogs*

Inland	Maritime
1. Domed	1. Plateau
2. Lagg (moat) poorly developed	2. Lagg often well developed
3. Surface pools common in center of bog	3. Surface pools rare
4. Trees often in center of bog	4. Trees, if present, restricted to slope of bog
5. Dwarf shrub communities cover most of bog	5. Dwarf shrub communities only on slopes and edge of plateau; deer's-hair sedge (*Scirpus cespitosus*) lawns on plateau
6. Huckleberry (*Gaylussacia*) community restricted to these bogs	6. Deer's-hair sedge–sphagnum and black crowberry (*Empetrum nigrum*) communities restricted to these bogs
7. Black crowberry, deer's-hair sedge uncommon; baked-apple (*Rubus chamaemorus*) absent	7. Black crowberry, deer's-hair sedge, and baked-apple berry abundant

SOURCE: Adapted from Damman (1977).

The moisture surplus—whether it is achieved by poor drainage, higher humidity, greater cloud cover, or "lock-up" from the cold—ensures that peatland plants, especially the main peat builders (sphagnum mosses and sedges) have what they need to thrive. They are able to do so well that peatlands build both vertically and horizontally. Under the most favorable conditions they can spread beyond the boundaries of basins, mound up into large but gradually rising domes, form elevated plateaus, and even disregard the terrain and climb slopes.

There are two types of raised bogs in the Northeast, differing in both their formations and their vegetation. Their distribution is strongly linked to the influences of oceanic and continental climates (see Table 5).

In the northern inland (continental) climate of the Northeast, raised bogs are domed—some imperceptibly, some noticeably. The surface slopes gently toward the edges of the bog

Fig. 21 *Maine's Great Heath*. This strikingly raised 4,000-acre coalesced domed bog has the convexity of an inland bog but is sufficiently close to the sea in eastern Maine to have several species more common on maritime peatlands. Dwarf shrubs, especially sheep laurel (*Kalmia angustifolia*) and leatherleaf (*Chamaedaphne calyculata*), dominate this view looking up the marginal slope to the central bog expanse with scattered ponds and isolated black spruce (*Picea mariana*). *(Photo by Hank Tyler.)*

(although the dome is not necessarily in the center of the bog). The lagg (moat) at the edges is either poorly developed or absent.

Many large inland raised bogs have pools or small ponds dotting the surface that were formed long after deposition of the peat. These features may be clues that the bog has reached a steady-state condition. Although they appear small, shallow, and temporary, some pools lie in place for hundreds or even thousands of years. An aerial view may show pools and/or vegetation arranged in a variety of pronounced patterns. The pools may be in concentric circles radiating out from the center of the peatland, like ripples from a pebble dropped in a pond. If the bogs are on gentle slopes, the pools

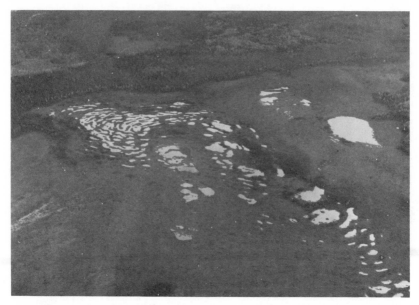

Fig. 22 *Patterned raised bog.* Crystal Bog in north central Maine has numerous surface ponds, many along concentric contours defining the higher central area. Other ponds may lie in wetter seeps along the bog surface. This aerial view includes the central 500 acres of this remarkable peatland. *(Photo courtesy of Maine Critical Areas Program.)*

may fan out from the top of the slope down or may be elongated into wide ladderlike tracks with strips of vegetation in between. In some cases the patterning may involve only the vegetation, as recurring assemblages of different plant types. The bogs often have "rays" of trees (usually black spruce) emanating from the highest point, with shrub or moss dominant throughout the remainder.

The plant communities of inland raised bogs are distinct and seem to follow gradual moisture-nutrient gradients from the border inward and upward to the dome, from moist and minerotrophic at the edges to relatively dry and ombrotrophic at the height of the dome (see the cross sections in Fig. 23–24).

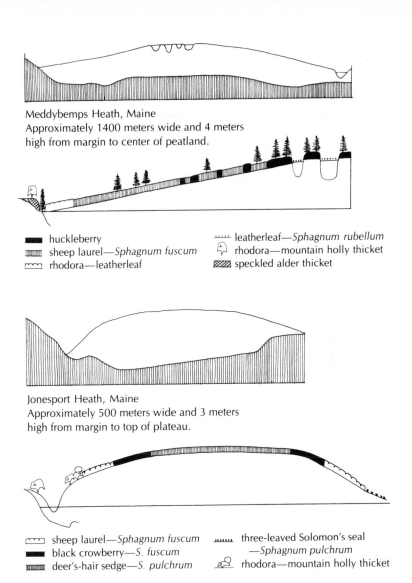

Meddybemps Heath, Maine
Approximately 1400 meters wide and 4 meters
high from margin to center of peatland.

- ▬ huckleberry
- ▦ sheep laurel—*Sphagnum fuscum*
- ▭ rhodora—leatherleaf
- ⋯ leatherleaf—*Sphagnum rubellum*
- rhodora—mountain holly thicket
- ▨ speckled alder thicket

Jonesport Heath, Maine
Approximately 500 meters wide and 3 meters
high from margin to top of plateau.

- ▭ sheep laurel—*Sphagnum fuscum*
- ▬ black crowberry—*S. fuscum*
- ▦ deer's-hair sedge—*S. pulchrum*
- ⊔⊔ three-leaved Solomon's seal
  —*Sphagnum pulchrum*
- rhodora—mountain holly thicket

Figs. 23 and 24 *Coastal plateau bog versus domed bog.* In eastern Maine rapidly changing climatic gradients produce dramatically different bog types only a few miles apart. The coastal plateau Jonesport Heath (now being mined to extinction) is nowhere more than a mile or so from tide-water. Domed Meddybemps Heath is but 15 miles from the Gulf of Maine yet strikingly different in overall topography, the presence of ponds, and plant communities. *(Adapted from Damman 1977.).*

Although they share many characteristics and species with inland raised bogs, most coastal raised bogs look like plateaus (hence the name "plateau bog"). They have a well-defined moat around the edge and rise steeply at first, then more gently, to a wide plain. The plain occupies most of the surface area of the bog and is nearly flat and nearly featureless. Unlike continental bogs, maritime bogs rarely have hummocks, hollows, or pools.

As in inland bogs, plant communities are distinct; the species and zoning, however, are different. The lagg is wet and minerotrophic, with such species as sweet gale (*Myrica gale*), three-leaved Solomon's seal (*Smilacina trifolia*), and nutrient-demanding sphagnum mosses. The shrub thickets of the lower slope of inland bogs are either less well defined or entirely absent. The steeper slope is similar in composition to the broad slope of the continental. But as the shoulder rounds off nearer the bog plain, black crowberry (*Empetrum nigrum*), rarely found inland, seems to take over from sheep laurel (*Kalmia angustifolia*), along with some more unusual species such as cloudberry (*Rubus chamaemorus*) and common juniper (*Juniperus communis*).

The upper shoulder and the plain are treeless. The center of the broad plain is usually wet in summer, whereas the steep slope is dry and well drained. In contrast, during the summer inland raised bogs are driest at the top of the dome and progressively wetter toward the edges.

It is curious that several species of plants growing in coastal raised bogs of the North Atlantic seaboard are either found only in alpine areas further inland or are restricted to fens in the southern Northeast or fenlike parts of continental raised bogs. Black crowberry, for example, in the Northeast lives mostly in scattered colonies on the high mountain summits of Maine, New Hampshire, Vermont, and New York. But in the maritime regions from Maine into Canada it is an abundant species, especially in plateau bogs.

One reason for this similarity may be that alpine and mari-

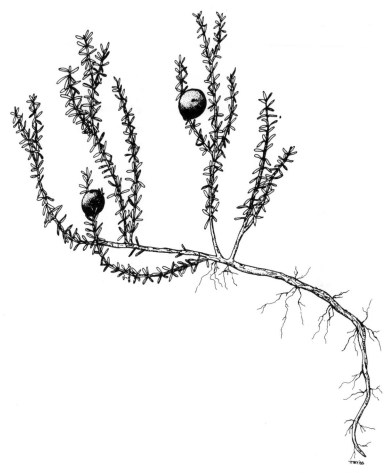

Fig. 25 *Black crowberry* (*Empetrum nigrum*). This sprawling, small-leaved shrub grows in profusion on coastal Maine bogs. Common at more northerly latitudes in New England and in New York, it also grows in many alpine areas and on eastern Maine coastal headlands. It has large black fruit (which makes tasty pies and preserves) arising from a small flower.

time areas have somewhat comparable climates. Both have their "heads in the clouds" a great deal (alpine areas literally in clouds, maritime areas in fog), both receive lots of precipitation, both have significantly lower evapotranspiration rates as a result of a shorter growing season (alpine) or greater overcast (maritime), and both have cool summers.

Also found in coastal raised bogs, along with their abundance of acid-tolerant boreal plants, are species that otherwise grow only farther south or inland: common juniper (*Juniperus communis*), a shrub of dry, rocky, or sandy soils; skunk cabbage (*Symplocarpus foetidus*), which belongs to riverbank communities, swamps, and fens of milder climates; and bayberry (*Myrica pensylvanica*), another inhabitant of sandy or rocky soils from Cape Cod south. The presence of such species may be due, at least in part, to the ombrotrophic bogs' proximity to the ocean. The bogs are close enough to get mineral enrichment from the sea, from spray or through fog or rain. This extra dose of salts may be enough to encourage the "outsiders" to stay, but other factors may also be influential, such as dry surface conditions in some parts of the raised bogs.

PEATLAND TOPOGRAPHIES

Like all other landscapes, peatlands have distinctive topographies. I have watched people take their first peatland steps: Initially hesitant and apprehensive, they soon squeal with delight at the feel of the quivering, spongy surface beneath their feet. Then, with surprise, comes the discovery that there is no smooth surface to walk on, but rather a dimpled, soggy land of ever-undulating hummocks and hollows. The densely vegetated mounds, typically 1 to 2 feet above the shallow, wetter depressions, may be firm and wooded or soft and delicately formed by loose-growing sphagnum. The hollows can be richly vegetated or filled with the decaying remains of nearby plants. Some are damp and some are wet; some of those that bear year-round water are truly pools.

Hummocks and hollows, along with vegetation and wetness, vary greatly from place to place in a peatland and from peatland to peatland. It is in walking amid them that one feels the distinctiveness, almost a personality, of each peatland type. Strong is their character, from the firm, dwarf shrub hummocks of a Maine raised bog to the soggy sedge and sphagnum hollows in a Pennsylvania fen.

One of the most active areas of peatland research seeks to understand why there should be these hummocks and hollows—how do they form and what functions do they serve? There are likely no simple answers, but several factors of importance are emerging. Certainly the plants play a major role, whether differential sphagnum growth, tussock-forming sedges, spreading shrubs, or colonizing mosses. In bogs, they may be determined largely by the growth habits of different sphagnum moss species (see Chapter 7).

Other forces also may accentuate, alter, or even suppress this surficial topography. Freeze-and-thaw cycles may intensify hummock patterns. Fire, not infrequent in peatlands, may burn higher and drier hummocks while passing harmlessly over saturated hollows. In large slope peatlands, gravity may control ridge and depression patterns through the downslope movement of water. In ribbed fens, differential decay on either side of the vegetated ribs may account for much of the patterning.

The ecological significance of hummocks and hollows may prove to be complex and profound. It appears that some hummocks and hollows are related to long-term means and extremes in water levels. Others may reflect variable rates of peat production and decay, perhaps the result of a fine balance between climatic and hydrologic conditions. If they are indeed keys to environmental conditions and changes, once we learn how to interpret their significance, hummocks and hollows will have an importance well beyond their actual size.

Thus, while at first meeting the surface topography may be little more than a curiously amusing inconvenience, it may be as important for the ecosystems of peatlands as fallen logs and leaf litter are to a forest, or dunes to a beach community, or grass tussocks to a prairie.

The uneven surface trod by bog-walkers is only the first of many variations in peatlands of the Northeast. There is quite a walk yet to come.

In the early stages of peatland formation, much of the direction of development is influenced by environmental factors: climate and geography combined with the controlling effects of water. But as the peats laboriously amass and build into thick mats, they become a factor in themselves, superimposing their influence on those from the outside. The evolving organism we call peatland, as an independently functioning system, takes on increasing control of its own destiny and becomes less dependent on the immediate surroundings. In their fullest expression in the raised bogs of the maritime regions, these systems are beholden to what falls from the sky for sustenance. This autonomy, splendidly unique, brings with it unusual biological relationships and elements, and we who observe gain our rewards.

5 | THE GEOGRAPHY OF NORTHEASTERN PEATLANDS

In the Northeast partial inventories and statewide estimates indicate that there may be as many as 20,000 individual peatlands, some no larger than a fraction of an acre, others expanses over a mile across. Most are small, less than 25 acres. A few are very large, more than 1,000 acres. A lot are in between, especially in the 25- to 250-acre range. In the southern part of the region, the peatlands tend to be isolated and small, though some do cover several hundred acres. But in the northern states peatlands become dramatically more prominent.

North America can be divided into twenty or thirty large biogeographic regions based on peatland type, from the High Arctic Region of basin and seepage fens over permafrost to the Southern Florida Region of eutrophic glades and swamps. Because of its location, the Northeastern United States is in several regions, thus providing its fascinating diversity of peatlands.

Southern New Jersey lies at the northern end of the Atlantic Coastal Plain region of grassy savannas, shrubby pocosins, wooded swamps, and coastal marshes. Northern coastal Maine belongs in the Atlantic boreal region of concentric, eccentric, and plateau raised bogs. The eastern deciduous forest region in the Northeast generally extends from southernmost Maine along south coastal New England, Long Island, and northern New Jersey through the southern two-thirds of Pennsylvania; however, the peatlands there are mostly mesotrophic swamps and shrub thickets, oligotrophic, open poor fens, and occasional herbaceous marshlands (by virtue of their species) which divide the area into coastal and interior subregions. Northern Pennsylvania and the northern mountainous areas of New Jersey, Connecticut, and Massachusetts, the bulk of New York, all of Vermont, and most of New Hampshire and Maine belong to a large region extending westward through the northern Great Lakes to the prairie boundary. Southern parts of this region have a variety of open and wooded fens and oligotrophic, nutrient-poor fens, whereas in the north of the region ombrotrophic raised bogs and patterned fens occur at their southern limit in boreal regions of North America.

To our north in Canada (and northernmost Minnesota) peatlands are larger, more complex, and less confined to basins. Grand and imposing landscapes, with distances measured in miles instead of yards and areas in square miles rather than acres, the northern muskeg can be an endless scenery, spreading as far as the eye can see—quite a contrast to the much smaller peatlands of the Northeast scattered as enchanting nuggets about the land.

Coolness and wetness, as noted earlier, can work hand in hand in the development of peatlands. Coolness can restrict evaporation, increase cloud or fog cover, and retard decay. Persistent wetness, whether the result of conservation or abundant supply, can decrease oxygenation and decay and provide for peat deposition. Thus, places with cool to cold northern climates, especially those that are humid and/or have relatively poorly drained substrates, are good for peatland development—hence the extensive peatlands in much of coastal and maritime Canada and the prominence of peatlands in the northern and coastal areas of the Northeast.

On the other hand, further south, wetness and warmth may team in peatland growth. In lowlands with persistently high water tables, extensive peatlands that are reminiscent of the great coal-producing peatlands of the Carboniferous period (345–280 million years ago) can develop. Humid, warm climates with long growing seasons permit abundant plant production, overloading still or slow-moving water with organic debris. Decay organisms deplete limited amounts of dissolved oxygen, and peat accumulates. Thus, on the poorly drained, nearly flat plains of the warm and humid Atlantic coast—from Florida's Everglades to Virginia's Dismal Swamp and southern New Jersey—peatlands form hundreds of thousands of acres of marshes, glades, swamps, and pocosins.

So the humid coastal climate, whether warm or cool, aided by poorly drained soils or topographic basins, permits abundant peatland formation. Such formation may be on elevated marine clays in nonglaciated southern New Jersey and glaciated eastern Maine or amid the dimpled, glacial recessional landscapes of southeastern Massachusetts.

There have been significant climatic changes during the last glaciation and even in the relatively brief period since the ice retreat (roughly 18,000 to 10,000 years ago in the Northeast). The climates that came with the glaciers south to Cape Cod, Long Island, and central Pennsylvania encouraged the development of peatlands along the cool and moist fringes beyond the glacier's limits, thousands of years before peatlands began to form to the north following the ice retreat.

When the climate warmed and the continental glacier retreated, vast tracts of barren lands became available again for recolonization by formerly dispossessed plant and animal species. Many of the Northeast's peatlands have their origins in the first or second millennia following deglaciation. Later, during the warmer and perhaps somewhat drier hypsithermal interval (from roughly 7,000– 8,000 to 3,000–4,000 years ago), other southern species crept northward, in many areas eventually replacing more northern communities. During this interval peatlands of the Northeast, especially in the southern sector, may have had difficulty in maintaining sufficient water balances in some marginally acceptable sites, thus accentuating their present restricted distribution (enhanced, of course, by the destruction of peatlands by drainage and flooding by man during the last 3½ centuries).

In many parts of the world, including the northern lake states and eastern Canada, with the somewhat cooler and wetter conditions following the close of the hypsithermal interval some 3,000 to 4,000 years ago, there was a resurgence of peatland growth and initiation; perhaps the Northeast's peatlands reacted similarly. In recent times our peatlands appear to have thrived in the climate of the last 3,000 years, but one wonders how they will fare in our contemporary, pollutant-rich, anthropologically altered climate.

The glaciated portion of the Northeast includes all six New England states, most of New York, the northwestern and northeastern corners of Pennsylvania, and the northern half of New Jersey (Fig. 26). Landscapes formed by melting and retreating glacial ice can provide abundant settings for peatlands. Trapped amid the ridges of terminal moraine systems (in southern New England, Long Island, and parts of New Jersey) are irregular depressions, often poorly drained due to high silt and clay content in the morainal till. Stagnant ice and meltwater features, such as kame-and-kettle topographies, eskers, outwash, and shallow ephemeral impoundments, create numerous opportunities for peatland development. Across northern New York (north of the Adirondacks and west central highlands) and in much of Maine these glacial deposits locally impede drainage, pond the water, and direct stream flow. Along coastal Maine, as ice retreated northward, the sea level rose more rapidly than the land could rebound from the depression caused by the great weight of glacial ice. Glacial marine deltas and subsequent clay-coated landscapes (formed as the land finally rose above sea level) provide even more peatland habitat.

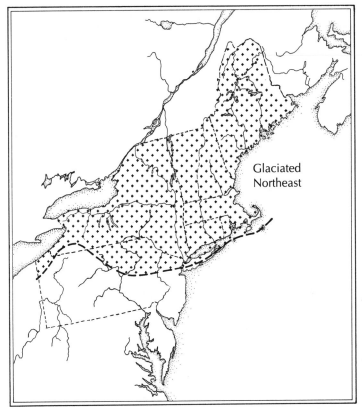

Fig. 26 *Extent of glaciation in the Northeast.* Glacial ice covered most but not all of the Northeast during the last (Wisconsin) glaciation.

Poor fens and essentially ombrotrophic level bogs are distributed throughout the glaciated Northeast. Although most of these over the years have been lumped together and referred to as "quaking" or "kettlehole" bogs, they are associated with a much greater diversity of geomorphic features than just quaking bogs with floating mats.

Those with "northern" complements of plants (see Table 4) occur in mountainous areas and northern lowlands—terrain generally forested with northern hardwoods such as beech (*Fagus grandifolia*), yellow birch (*Betula lutea*), and sugar

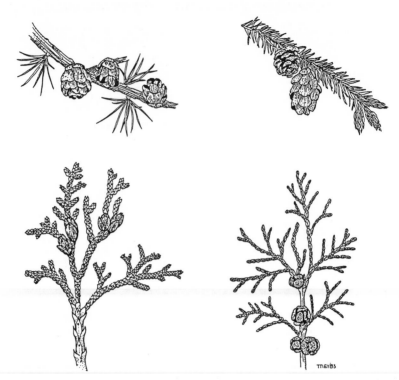

Fig. 27 *Peatland conifers.* These and a few other conifers grow on north-eastern peatlands, each species displaying geographical restrictions and habitat preferences. *Top left*: Tamarack or eastern larch (*Larix laricina*). Mostly northern in our region, its presence indicates slight to moderate nutrient enrichment. It is the only native conifer of the Northeast that sheds its needles in the fall. *Top right*: Black spruce (*Picea mariana*) occurs in bogs and nutrient-poor oligotrophic conditions. The dark, small cones remain attached to branches for several years. Its tall, spindly shape is distinctive. *Bottom left*: Northern white cedar or arbor-vitae (*Thuja occidentalis*), an indicator of calcium-rich waters, predominates in forested fens in the northern tier of states. It also grows in pure stands in swamps, often in association with rare orchids. Most cedar swamps, peatland or otherwise, are becoming increasingly scarce as a result of cutting, draining, and filling. *Bottom right*: Atlantic (or southern) white cedar (*Chamaecyparis thyoides*), a Coastal Plain cedar, is equally at home in peatlands and other swamps. Scarce in southern Maine and New Hampshire, it becomes more frequent farther south. Long prized for shingles, posts, and other products, Atlantic white cedar has been cut extensively and even has been exhumed from burial in peat for use in regions where no standing stock remains.

maple (*Acer saccharum*) or trees of the boreal forests, notably red spruce (*Picea rubens*) and balsam fir (*Abies balsamea*). Even if we lift our gaze to the highest peaks and ridges of the Northeast, we find such peatlands. The alpine reaches—the lands above treeline—have some small oligotrophic bogs in bedrock depressions or more spreading systems of poor fens on slopes. These alpine peatlands have many of the same plants as their oligotrophic cousins at lower elevations. The slopes and ridges of Mount Mansfield in Vermont, Mount Washington in New Hampshire, and the Mahoosuc range in Maine all have small peatlands of this type.

In the more rolling and flatter lowlands of southern New England, New York, Pennsylvania, and New Jersey, where forests contain many oaks (*Quercus* species) and pines (*Pinus* species), the peatland vegetation has a more southern character.

Clearly, the glaciers' physical remodeling of northeastern landscapes had a direct bearing on the development of peatlands. But the effects of glacial activity also persist in other equally important but more subtle ways. Stretching from southeastern Massachusetts and Long Island, down the Atlantic seaboard, around the tip of Florida and along the Gulf of Mexico coast is the great flatness of the Atlantic and Gulf Coastal Plain. This large expanse, in some places as wide as three hundred miles, gently slopes from interior uplands to the sea. At various times in the distant geologic past the Coastal Plain was beneath the ocean, receiving sediments eroded from the neighboring higher ground. At intervals these ancient continental shelves uplifted, exposing sedimentary deposits as wide flats of sand or clay, only to be submerged by another underwater episode. Thus, events occurring over a period of 150 million years resulted in massive depositions of sediments, in some places more than a mile thick. These sediments typically are poorly drained and of low relief, thus favoring peatland development.

Then, more recently, during the Pleistocene, the Coastal

Plain was further influenced by the global advances and re-treats of glacial ice. For although the Pleistocene glaciers did not reach farther south than Long Island, at the height of the last glaciation (some 18,000 years ago), with tremendous quantities of water bound up as ice in glaciers, the Atlantic Ocean was as much as 330 feet lower than it is today. Conse-quently, the present-day continental shelf was exposed several miles to the east, expanding the Coastal Plain by thousands of square miles and providing additional habitat (no doubt including peatlands) for plant and animal species excluded by ice to the north. Furthermore, the glacially cooled climate ex-tended southward over all of the unglaciated Northeast, per-haps enhancing peatland development in this region and cer-tainly promoting vigorous colonization by boreal species.

As the climate gradually warmed and the glaciers retreated, the sea level rose, gradually reclaiming the most recently ex-posed shelf. But these lowlands, newly formed and dynamic barrier beaches, and the youthful glacial landscapes served as highways for the migration of southern plants northward. Be-fore the close of the hypsithermal interval about 3,000 years ago the climate ameliorated so much that some southern plants such as black gum (*Nyssa sylvatica*) made their way up to northern New England. Although the climate has cooled somewhat, a few of the species continue to grow in isolated northern locations in relic populations. Black gum, for ex-ample, has persisted in a northwestern Vermont peatland bor-dering Lake Champlain. The famous Pine Barrens of New Jersey constitute an especially fine meeting ground: there more than 100 southern species reach their northern limit, and about fifteen more species are at their southern limit. A number of these grow in peatlands (see page 67, for example).

Northern New England and New York have extensive gla-cial recessional landscapes (especially where large outwashes, eskers, and ephemerally inundated lowlands created poorly drained topography of low relief). These, combined with cli-

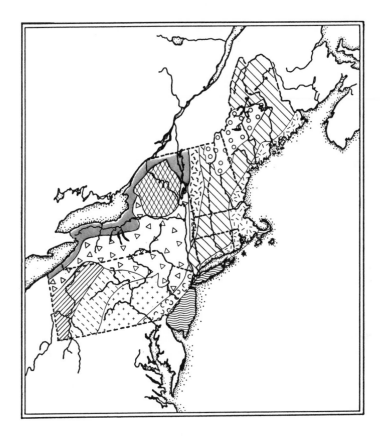

Fig. 28 *Physiographic provinces.* These major physiographic divisions of the Northeast are based on differences in landforms, geologic history, and climate. Many species of both animals and plants have distributions reflecting these distinctive regions. *(Adapted from Fenneman 1938.)*

mates that provide nearly as much or more precipitation per year as is lost by evaporation and transpiration, account for some of the Northeast's most fascinating, most unusual, and largest peatlands.

Ribbed fens cover tremendous acreages of central and northern boreal Canada, Alaska, Fennoscandia, and western Siberia. They reach their southernmost North American limit in the northern fringe of New York's Adirondacks (two peatlands with local rib development) and northern interior Maine (several peatlands with excellent rib formation). Only very recently discovered, the Northeast's ribbed fens occupy shallow slopes and appear to have abundant water for much of the growing season.

Likewise, this area is the only known location in the United States (excluding Alaska) of large, prominently raised, convexly domed bogs. In the Northeast these are at their southern extreme in North America. Apparently always some distance from the sea, these raised peatlands differ from the less clearly raised bogs of even more inland areas (e.g., northward from northern Minnesota) in their openness and concentric and eccentric patterning, in contrast to the generally wooded radial design of the most continental (inland) types.

This area also is the only known location in the United States outside of Alaska to have coastal plateau bogs. In limited and decreasing number (due to mining and other intrusions) along Maine's easternmost coast, these flat-topped, raised bogs are never more than several hundred yards from tidewater. At their southern limit in North America in the Northeast, coastal plateau peatlands occur eastward and northward along coasts to western Newfoundland and nearby shores of the Gulf of St. Lawrence.

Maine and northern New York possess many large peatlands (500 to 4,000 acres), from sedge fens to dwarf shrub and sphagnum ombrotrophic bogs. The absence of equally large peatlands in northern Vermont and New Hampshire seems primarily due to the limited extent of glacial outwash

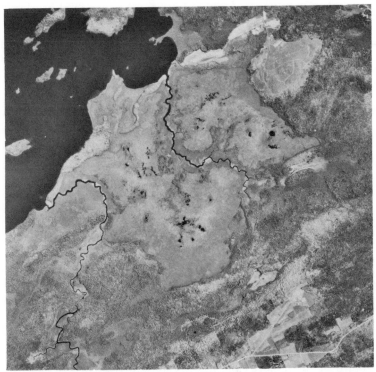

Fig. 29 *Meddybemps Heath, Maine.* The aerial photograph shows the 2,000-acre coalesced domed peatland, with its complex of streams, ponds, and vegetation types. *(Photo courtesy of Maine Critical Areas Program.)*

and stagnant-ice topography rather than to climatic differences. The large New York peatlands (e.g., Massawepie Fen) may owe their size mostly to the existence of large shallow basins, whereas in eastern Maine—in the region of prominently domed bogs—lateral expansion (with accompanying paludification) has extended peatlands beyond their originating basins, sometimes crossing shallow divides and occasionally causing the coalescence of two or more peatlands into a large, complex, undulating bog (e.g., Maine's Great Heath).

Along the Gulf of Maine from Acadia National Park east

into New Brunswick there is another type of special peatland. It resembles the most extreme expression of oceanic peatlands, the so-called blanket bog. A true blanket bog knows few bounds, cloaking the land across valleys, up slopes, and over ridges. It dominates the landscape, just as do forests in northern New England and open farmlands in the Midwest. Blanket bogs are found on foggy, cool-temperature coasts, notably in Ireland, Scotland, and the southern tip of South America. They also occur on the west (in British Columbia and Alaska) and east coasts of North America. In all cases, annual precipitation exceeds 50 inches and is relatively evenly distributed throughout the year; fog is common, especially during summers.

No true blanket bogs have been found in the Northeast (southeastern Newfoundland contains eastern North America's only extensive true blanket bog). However, thin peat soils, supporting dwarf shrubs, sedges, and sphagnum, blankets undulating land in a few locations. On headlands of Maine's damp eastern coast, heathlands with 6- to 24-inch-deep peat soils (sometimes called "dry" peatlands, since through much of the year there is no standing groundwater in them) persist where rain and fog mitigate potential droughts.

Along the Northeast's coast, tidal marshes lie behind barrier beaches, fill lagoons and embayments, and border the mouths of rivers and streams. Many contain peat or peat-rich horizons and soils. The peats in the upper horizons are usually little decomposed; more than 70 percent of the plant fibers can be identified even hundreds of years after burial. Now under as much as 30 or more feet, freshwater peats (typically with much sedge material) underlie some marshes, having formed shortly after the departure of the glaciers but prior to inundation by the rising sea.

The plants in a number of the marshes grow on and contribute to peats and organic-rich soils. Before European colonization and clearing of the erosion-protecting upland forests—and before the advent of mechanized construction work—some of these peat soils apparently had lesser proportions of mineral sediments. On these wet soils common plants include salt meadow grasses (*Spartina patens* and *S. alternifolia*) and spikegrass (*Distichlis spicata*). There also may be the rush "blackgrass" (*Juncus gerardi*), the succulent saltwort (*Salicornia europaea*), sea lavender (*Limonium carolinianum*), and colorful flowers such as seaside goldenrod (*Solidago sempervirens*). In estuarine marshes, where river water dilutes the marine salt content, one may find narrow-leaved cattail (*Typha angustifolia*), the giant reed *Phragmites communis*, and the bulrush *Scirpus robustus*.

The productivity in tidal marshes ranges from moderate to high, with some estuarine systems producing more plant and animal material per unit area per year than most or all other ecosystems in the region. This, of course, is in sharp contrast to the extremely low productivity of oligotrophic and raised bogs (Table 6). Tidal wetlands are well known for their importance to a great many species of fish and birds, which, in turn, are a direct indicator of the productivity of such places.

The productivity of wetlands—how much living "stuff" they can make every year—is one of the most variable of ecological measurements (see Table 6). At one extreme are marshes, with some of the most productive of all ecosystems on earth, surpassing even the most fertile agricultural lands or lush forests. At the other extreme, ombrotrophic bogs are some of the poorest producers, with only the arctic tundra (very cold deserts), some dwarf scrub-shrub deserts (very hot deserts), open ocean, and some lakes and streams poorer.

Man's alteration of coastal lands and waters has greatly changed the shape and character of most tidal wetlands. Diking, channeling, draining, rail and road building, and other momentous restructuring have left long-lasting scars. Changes in water chemistry, flooding periodicity, silt and sand deposition, and activities such as haying and burning have much altered plant and animal communities. As one reads the ancient past in diverse sediments thousands of years old, one realizes

TABLE 6   *Generalized Primary Productivity of the Earth's Major Ecosystems*

Ecosystem	Annual Productivity of Biomass by Green Plants (grams per square meter)
Tropical forests	2,200
Swamps and marshes	2,000
Northern temperate forests	1,250
Boreal forests	800
Shrublands	700
Peatlands	300*
Lakes and streams	250
Open ocean	125
Deserts	90

SOURCES: Whittaker (1975); Worley (1981).
* An approximate average; peatlands range from less than 100 g/m² to near 2,000 g/m².

that these alterations often equal, sometimes exceed, the changes wrought by nature: shifting climates, meandering rivers, migrating species, and wandering water levels.

Thus, the length and breadth of the Northeast is rich with kinds of peatlands—far richer than was suspected even a short time ago. From the larger regional perspective, let us now focus briefly on the peatlands of the individual states.

NEW JERSEY

The third smallest northeastern state, with 7,836 square miles, New Jersey has a surprising variety of landscapes. On either side of a line extending from southwest to northeast at the narrow "neck" of New Jersey, the scenery is dramatically different. Northern New Jersey, subjected to three of the four major ice advances during the Pleistocene, has the rolling topography typical of the Allegheny foothills and the Piedmont. Unglaciated, low-lying southern New Jersey is Atlantic Coastal Plain—a terraced, nearly flat ancient sea bottom of marine sands and clays.

Peatlands of these two regions, though generally different, share many species. Widespread are heaths, including leatherleaf (*Chamaedaphne calyculata*), large cranberry (*Vaccinium macrocarpon*), highbush blueberry (*Vaccinium corymbosum*), black huckleberry (*Gaylussacia baccata*), and sheep laurel (*Kalmia angustifolia*), as well as several more southern plants such as swamp azalea (*Rhododendron viscoscum*), sweet pepperbush (*Clethra alnifolia*), staggerbush (*Lyonica mariana*), and inkberry (*Ilex glabra*). Other plants common to both are several sphagnum species, pitcher plant (*Sarracenia purpurea*), sundews (*Drosera* species), Virginia chain fern (*Woodwardia virginica*), and numerous sedges.

The peatlands of northern New Jersey often contain black spruce (*Picea mariana*), tamarack (*Larix laricina*), Labrador tea (*Ledum groenlandicum*), and bog rosemary (*Andromeda glaucophylla*)—species not found in the sand plain bogs. On the other hand, the Coastal Plain has some species absent or rare in the northern highlands. Atlantic white cedar (*Chamaecyparis thyoides*), sparse in the north, forms very dense and pure stands (though in many places it is heavily logged), under which may grow such species as sweet bay (*Magnolia virginiana*), wax myrtle (*Myrica cerifera*), bog asphodel (*Narthecium americanum*), and curly grass (*Schizaea pusilla*), plants unknown in the more northern peatlands.

In general, the Coastal Plain peatlands are shallower than their northern counterparts. Peatlands of both regions tend to be acidic, although the Coastal Plain ones—occupying sites with high water tables, springs, or adjacent streams—as a group are more nutrient-rich.

The million-acre Pine Barrens in the southern half of New Jersey is perhaps the best and most fascinating place in the Northeast to see Coastal Plain peatlands. The Cape May peninsula also has several fine examples, especially Atlantic white cedar swamps and peat-filled, herbaceously covered "Carolina bays." The Pine Barren peatlands have not only an unusual biology, with many endemic and disjunct plants and

Fig. 30 *New Jersey peatland*. A Pine Barrens pond with rich bog vegetation growing over its surface. *(Photo by Robert A. Zampella.)*

animals, but also an interrelated history of commercial use. Bog ore (iron-rich precipitates in lake, pond, and stream bottoms now covered with peat) was mined, then smelted nearby to make iron, especially for foundries. Atlantic white cedar was much cut for posts, railway ties, and building materials. Many sites were modified for cranberry harvest. Some of these activities continue today; others can be learned about further at historic sites open to the public.

PENNSYLVANIA

Peatlands are widely distributed throughout Pennsylvania, the Northeast's second-largest state at 45,068 square miles, although nowhere are they a prominent landform. They lie astride shallow divides, fill abandoned stream courses and floodplain depressions, and persist 'here springs or seeps saturate gentle slopes. In the glaciate northeastern and northwestern corners of the state small oligotrophic bogs occupy kettleholes in glacial deposits. Most of Pennsylvania's extant peatlands cover less than 100 acres, but a few exceed 250 acres. Some are part of much larger wetland complexes.

Poor fens—some open, others thickly wooded with shrubs or trees—are probably the most frequent peatland type, although oligotrophic sites (a few with floating mats and central ponds) may be more easily recognized by a greater number of people. Minerotrophy commonly results from ground or surface water coming in contact with the widespread sandstones and limy shales of the Allegheny Plateau and the Ridge and Valley province.

As in much of the Northeast, many of Pennsylvania's peatlands were caused by, or at least heavily influenced by, beavers (see Chapter 14). Since European colonization in Pennsylvania, beaver colonies (hence many related peatlands) have had to contend with much human disturbance, especially periodic elimination of their forest homes, trapping, and dam destruction. Surely many such peatlands become farmland.

The plants of Pennsylvania peatlands have both northern and southern affinities. Although some of the boreal heaths (notably leatherleaf *Chamaedaphne calyculata*) are locally abundant, others such as rhodora (*Rhododendron canadense*), Labrador tea (*Ledum groenlandicum*), and bog laurel (*Kalmia polifolia*) become less frequent and more site-specific. Shrub thickets, often of highbush blueberry (*Vaccinium corymbosum*) and black highbush blueberry (*V. atrococcum*) disguise many peatlands for persons looking for open "bogs." Rhododendron (*Rhododendron maximum*) can be prolific at the peatland-upland borders, making passage most difficult and irksome. Where peatlands intergrade with hardwood swamps, especially in river bottomlands at lower elevations in the southern and southeastern part of the state, one may have poison ivy (*Toxicodendron radicans*), poison sumac (*Toxico-*

Fig. 31 *Pennsylvania peatland.* Tangles of rhododendron (*Rhododendron maximum*) surround a poor fen in east-central Pennsylvania. Cinnamon fern (*Osmunda cinnamomea*), seen behind the figure, is part of a lush herbaceous understory. *(Photo by Charles W. Johnson.)*

dendron vernix), tangles of greenbrier (*Smilax rotundifolia*), and deep mucky soils.

Scattered, vegetationally diverse, often wooded in shrubs or trees, and typically small, Pennsylvania's peatlands are slowly becoming better known. With time it is the hope of Pennsylvania conservationists that representatives of each type of the state's peatlands be placed in protective ownership.

CONNECTICUT

Connecticut, second-smallest of the northeastern states at 5,009 square miles, can be divided into several floristic provinces or physiographic regions, with four main areas: the central broad valley of the Connecticut River, the forested uplands and plateaus on either side of that valley, the flat seaboard, and the northwestern hills and valleys of the Taconic Mountains. The peatlands of Connecticut resemble those of adjacent Massachusetts and Rhode Island, their character determined in large part by coastal or glacial geomorphology combined with climatic variation afforded by proximity to the sea.

In Connecticut the term "bog" most commonly is applied to the most nutrient-poor sites, either open assemblages of sphagna, sedges, heaths such as leatherleaf, insectivores including pitcher plant and round-leaved sundew (*Drosera rotundifolia*), or wooded with black spruce, tamarack, or, to the south, Atlantic white cedar. Many developed in kettleholes; others are between drumlins, occupy bays of ponds, or lie in poorly drained basins at broad watershed divides.

In northwestern Connecticut a few richer fens depend on waters from calcium-based marbles and limestone bedrock. So nourished, these peatlands are the only ones in the state where northern white cedar (*Thuja occidentalis*) grows, along with a special and unusual assortment of other plants, including dwarf birch (*Betula pumila*), autumn willow (*Salix serissima*), and marsh bellflower (*Campanula uliginosa*).

Connecticut's peatlands often are part of larger wetlands. To the south one may find on peatland (or in wetland bordering peatland) Atlantic white cedar, sweet pepperbush, spicebush, winterberry (*Ilex verticillata*), poison sumac, mountain laurel (*Kalmia latifolia*), cinnamon fern (*Osmunda cinnamomea*), arrowwood (*Viburnum recognitum*), and other shrubs and herbaceous plants. Where there is the least enrichment from ground or surface water, sphagnum hummocks provide support for stunted leatherleaf, sheep laurel (*Kalmia angustifolia*), and small cranberry (*Vaccinium oxycoccus*). Richer areas may contain sweet gale (*Myrica gale*) and highbush blueberry, often in abundance. Pitcher plant is common. Some of Connecticut's numerous tidal marshes—both coastal and estuarine—have peat or organic-rich soils. The basal

Fig. 32 *Connecticut peatland.* Atlantic white cedar (*Chamaecyparis thyoides*) borders an open mat of low heath shrubs and sphagnum at this coastal bog. *(Photo by Charles W. Johnson.)*

sediments of a few deep coastal marshes (e.g., Milford Marsh) have a layer formed in freshwater marshes shortly after the melting of the glacier. Contemporary plant growth in the saline waters includes the salt marsh grasses (*Spartina* species), the common bladegrass rush, saltwort, seaside plantain (*Plantago juncoides*); in the less salty estuarine brackish water are rich growths of cattails, bulrushes (*Scirpus* species), or giant reed.

A few of Connecticut's peatlands are in protective management, though a burgeoning population in this long-settled state greatly threatens most remaining sites.

RHODE ISLAND

Almost all of Rhode Island, the nation's smallest state, with only 1,214 square miles (New York state is over 43 times as large!), is in the New England Seaboard physiographic province, with only the extreme northwest corner having elevated hills of the New England Upland province. As in neighboring Massachusetts, its oligotrophic and minerotrophic peatlands tend to be small, formed from kettlehole ponds, in lakeside bays, or at low-lying sites with high water tables. The several thousand acres of tidal marshes behind Block Island Sound beaches and around Narragansett Bay have peat-rich soils in a number of places.

The floral composition of the inland peatlands changes slightly north to south. Black spruce, for example, grows in some peatlands in the northern part of the state, whereas Atlantic white cedar is restricted to near the coast.

Distributed sporadically throughout the small state, open and shrubby poor fens are perhaps the best-known and protected of Rhode Island's peatlands. Typical for south coastal New England, shrubs such as sweet gale and bog rosemary and a variety of sedges are well represented.

Peatlands in Rhode Island are particularly vulnerable to overvisitation because the state is so small, the sites few, and the population relatively high.

MASSACHUSETTS

Although a fairly small state (its 8,257 square miles ranks it sixth of the Northeast's nine states), Massachusetts has a fascinating diversity of landscapes: the Berkshire Mountains in the extreme west, the central rolling uplands (bisected by the broad valley of the Connecticut River), the gently inclined eastern seaboard so richly mantled by glacial debris, and the sea-dominated Cape Cod and islands (Nantucket and Martha's Vineyard the best known) with their famous sands. Peatland abundance and variety in Massachusetts corresponds with this landscape spectrum, especially as it combines with climatic differences due to elevation and distance from the Atlantic. Several sites are well known, for they have quaking, floating mats around central ponds. One is even legendary for its literary heritage—Thoreau's Bog!

The lumpy, pitted-plain topography of the eastern coastal plain, including Cape Cod and the islands, owes its immediate texture to complexes of terminal moraines and outwash plains where the last retreating glacier dumped enormous quantities of sands, gravels, boulders, and clays. Beneath these often porous and well-drained materials are the compacted tills from previous glaciations and other poorly drained substrates from a more distant past. These kettlehole bogs or poor fens literally dot many, many areas. Long ago converted (via ditching, the addition of sand, and periodic flooding) to cranberry production, many form the basis of a large and thriving industry. The precolonial character of these peatlands is essentially unknown. Interestingly, though, locally the term *bog* refers also to artificial nonpeatland wetlands—some created from other wetlands such as red maple swamps, and still others in low areas or even abandoned gravel pits.

Poor fens in eastern Massachusetts occur at springs, by streams, or at the margins of slightly marly ponds and lakes. Some estuarine and tidal marshes also have deposits of peat. Fenway Park in Boston, famous as the home of the Red Sox, originally was part of a large wetland—"the Fens"—before most of it was filled during urban development. Some of these fens still exist, though converted to community gardens.

Atlantic white cedar peatlands, once common along the coast and out on the Cape Cod peninsula, have been repeatedly and extensively logged for their timber. Fortunately, a few have been preserved as natural areas. The cedar wetlands have various blends of northern species (such as mountain holly *Nemopanthus mucronata* and Labrador tea) and southern (sweet pepperbush, Virginia chain fern). On the Cape and islands curious little "dune bogs"—pockets of sphagnum

Fig. 33 *Massachusetts cedar swamp.* A boardwalk winds through this Atlantic white cedar (*Chamaecyparis thyoides*) swamp at Cape Cod National Seashore, providing visitor access while protecting the growing plants and surface peats. *(Photo by Charles W. Johnson.)*

mosses, sedges, cranberries, and other plants—nestle in between coastal sand dunes.

Across the state's north-central plateau and into the mountains, the peatlands have a more northern character. Peatlands along lakes or sluggish streams may be swamps with red maple, thickets of alder, or marshes thick with cattails or other narrow and broad-leaved herbs. These so-called bogs are nutrient-poor oligotrophic sites with boreal species, many of which are also residents on truly ombrotrophic bogs: black spruce, heaths such as leatherleaf and bog laurel, insectivores including pitcher plant and sundews, cotton grasses, and small cranberry. Sphagnum-rich, these sites border ponds, fill kettles and bedrock basins, and lie astride shallow divides. Few exceed 100 acres. A few nutrient-rich fens exist in the limestone belt of the Berkshires, in the western portions of the state.

Recently, active conservationists in state and federal government and especially in private organizations have identified and protected several rare, representative, and threatened peatlands in all sections of Massachusetts, from the densely populated east to the more rural western mountains.

NEW YORK

At 53,206 square miles, New York is by far the Northeast's largest state—in fact, 12,000 square miles larger than the six smaller northeastern states combined. Over 400 miles from Montauk Point, Long Island, in the east to Lake Erie in the west, and over 200 miles from Long Island's Atlantic beaches to the St. Lawrence River and Canada to the north, New York spans greater distances than any other northeastern state. Stretching through many climates, with mountain ranges, glacial topographies, major rivers, and lake and sea coasts, New York habitats are so diverse the peatlands are among the most varied in the Northeast. New York also has a lot of peatlands; a recent inventory listed more than 870 yet did not

Fig. 34 *New York peatland.* Deep in the Adirondacks, Spring Pond Bog displays some of its special features, as seen from the air. Here one of several shallow ponds, this the largest at about an acre, sits within the peat mat. Black spruce trees are evident; the lines indicate well-worn deer trails over the peatland. *(Photo by Ian Worley.)*

count the hundreds of sites containing only a few acres nor those of the 1,500,000-acre Adirondack Preserve.

The few but highly significant raised peatlands of the Adirondack Preserve in northeastern New York include some of the larger peatlands of the Northeast. Although to the untrained eye only imperceptibly elevated, they are the only clearly raised bogs in the Northeast outside of Maine. The two largest are quite different. One (crossed by a railway fill and bisected by a former lettuce farm) has a very dry surface with prominent sturdy hummocks of low shrubs and lichens;

the other is more typical, with scattered black spruce, a few surface pools, and a hint of concentric patterning. The latter has, along with one additional nearby site, the only ribbed fens in the Northeast other than those in northern Maine; these are the southernmost ribbed fens known in North America except for one site near Milwaukee.

Throughout the Adirondacks ombrotrophic (or nearly so), oligotrophic, and some slightly richer sites occupy kettles, spread across outwash flats and some narrow valleys, fill former lakes and lake bays, border streams and ponds, develop in beaver impondments, and generally form a characteristic part of this montane landscape. Most are less than 200 acres, but the sedges and sphagna, very wet, form a virtual sea of waving green (Massawepie Fen spreads over 900 acres or so).

The Great Lakes province of New York has many peatlands. The area was subjected to several glaciations, which have left their mark in the form of till, drumlins, blocked drainages, and sandy outwash plains. The strong limestone influence from the bedrock there has helped to create marly ponds, fens, and rich marshes. The fens are especially interesting: besides possessing the normal array of typical Northeast fen species, they also have some whose main range is more midwestern and that do not occur elsewhere in our region. Other unusual peatlands here are the interdrumlin poor fens, many of which exist in close association with larger wetland complexes.

Much of the southern half of New York is in the Glaciated Allegheny Plateau province, where peatland variety is great—from quaking acidic kettleholes to scattered open-basin poor fens to an occasional rich fen where limestone is present (poor fens are associated more with sandstones and shales).

Long Island's glacial origins and coastal dynamics provide a number of wet depressions containing ponds, marshes, and peatlands. Perhaps the most interesting in this Atlantic Coastal Plain topography are small and shallow peatlands tucked in low depressions between sand dunes where the water table is high and fairly stable. These "dune bogs" are really poor fens (with the influx of windblown sand, some intergrade with marshes), often well endowed with sphagnum mosses and sedges. As ecosystems, they have species and dynamics similar to the dune bogs on Cape Cod to the north and along beaches (where not heavily developed) of the mid-Atlantic states. Some interesting plants live in the dune wetlands, including thread-leaved sundew (*Drosera filiformis*).

In a state as large as New York human disturbance and the preservation of wildlands vary greatly from region to region. Along the great transportation corridor to the Great Lakes—up the Hudson River Valley then west to Lake Ontario and Buffalo—agricultural and urban expansion have destroyed or transformed many a peatland. But in the more mountainous lands, especially in the Adirondacks and less populated northern counties, many peatlands remain essentially as they were. Few of these are in protective ownership, though. Recognizing the importance of preserving rare and representative wetlands, some governmental agencies and private groups such as The Nature Conservancy have established local preserves and are seeking to secure other sites throughout the state. Current projects include protection in the northern Adirondacks of some of New York's largest and most significant peatlands.

VERMONT

Vermont's peatlands range in size from less than an acre to more than 200 acres, the larger primarily in the northern third of this mountainous state. A large proportion border lakes and ponds or combine with marshes and swamps to form wetland complexes. Others lie on shallow watershed divides, occupy bedrock basins in the hills and mountains, fill a few of the occasional kettles to be found in Vermont, and take advantage of impoundments created by rivers in the formation of deltas and the abandonment of oxbow meanders.

Fig. 35 *Vermont peatland.* A bog in Vermont reflects the solitude of the un-developed northlands. The floating mat supports a combination of sedges, low heaths, and sphagnum. Black spruce dominates the wooded back-ground. *(Photo by Marc DesMeules.)*

Open, shrubby, and wooded poor and extremely poor fens predominate, with richer sites most common in the southern marble belt and the east- and west-central limestone areas. There are no tidal peatlands, as Vermont is the only New England state with no Atlantic coastline. But near the top of 4,393-foot-high Mount Mansfield (Vermont's highest summit) two tiny windswept peatlands nestle in slight bedrock catch basins surrounded by alpine tundra.

Topography and climate account for the small to intermediate size of Vermont's peatlands. The mountains limit most peatland formation to rather confined valleys or small bedrock basins. Moreover, with the north-south trend of the mountains (notably the Green Mountains, which extend from Canada to Massachusetts) glacial passage and retreat left few deposits suitable for the development of large peatlands. Only the broad plain of the Champlain Valley of northwestern Vermont provides a variety of lowlands in excess of 100 acres suitable for peatland, yet even here peatlands of this size are scarce.

With the southernmost extent of raised bogs reaching the northern Adirondacks and well south in Maine, raised bogs should be expected in northern Vermont (and neighboring New Hampshire), but none is known. Nevertheless, a few of the state's larger northern peatlands (e.g., Franklin and Peacham Bogs) appear on the verge of being raised, and with precision survey they might in fact actually prove to have slightly elevated areas. They do share with the Northeast's ombrotrophic bogs similar vegetation, including hummocks with *Sphagnum fuscum* and even lichens, layered (the rooting of lower branches) and stunted black spruce, and the local absence of enrichment indicators such as bog rosemary (*Andromeda glaucophylla*). Although there are no hints of surface patterning, incipient surface pools can be recognized. Perhaps where climates are marginally acceptable for raised bog formation, there is a threshold size for the initial peatland, for none of Vermont's northern peatlands is even half as large as the nearest raised bogs in New York, Quebec, or Maine.

In northern Vermont, and throughout the state in the foothills and mountains, species common in boreal North America dominate the more nutrient-poor peatlands. Floating mats (quite uncommon in Vermont) and hummocky, open communities have numerous sphagna and sedges, heaths such as leatherleaf, sheep laurel, bog laurel, and Labrador tea, the creeping small cranberry, and, typically, abundant pitcher plant. Often subdued, black spruce and tamarack may be scattered throughout, form marginal wooded zones, or cover entire peatlands. At somewhat richer sites, especially where there is stream, lake, or beaver flooding, there can be much speckled alder, red maple, and even gray birch (*Betula populifolia*).

At lower elevations (especially in the milder Champlain

Valley) and lower latitudes (particularly where limestone, limy tills, and glacial lake clays predominate), densely vegetated peatlands of striking appearance have formed. Extensive northern white cedar swamps now are forested mostly by red maple, as the cedar has been nearly logged out. In places highbush blueberry rises in dense thickets up to 15 feet tall. Marshlands of cattails and carrs of alders form lake-margin wetlands over peat or peat-rich deposits. At some even richer, sedge-dominated fens uncommon species and colorful orchids (e.g., brook lobelia *Lobelia kalmii* and showy lady's slipper *Cypripedium reginae*) attract many naturalists.

Some southern species extend northward into Vermont at these peatlands—for example, shrubby cinquefoil (*Potentilla fruticosa*), maleberry (*Lyonia ligustrina*), and even black gum (*Nyssa sylvatica*). The last, although not a plant of peat soils in Vermont, has its northern limit lining the very margin of Colchester Bog on the east side of Lake Champlain.

Compared with those of southern New England, Vermont's peatlands are little disturbed, yet most show some trace of human activity. Commercial use is limited to the local sales of crudely dug peat for soil improvement. Public and private preservation activities protect a number of sites although by no means a fully representative sample of the peatlands throughout the state.

NEW HAMPSHIRE

New Hampshire's mountains, granite bedrock, northern New England climate, glacial history, and limited coastline largely determine the nature of its peatlands. Mountain landscapes (including the Northeast's highest summit, Mount Washington, at 6,288 feet elevation) dominate from Quebec to Massachusetts. Southeastern New Hampshire is a somewhat hilly coastal plain blanketed with sandy-gravelly glacial deposits. Sandy beaches line part of the Atlantic coastline. New Hampshire (9,304 square miles) and slightly larger next-door-neighbor Vermont (9,609 square miles) are the largest of the six smallest states of the Northeast.

Climatic variation is accentuated by this topography. The southeastern lowlands, though along the humid coast, are subjected to occasional very warm summer droughts. The central mountains are cooler from their height, alternately receiving the vagaries of continental air masses from the west and moisture-laden Atlantic storms from the south and southeast. The northern highlands have distinctly northern New England cool-moist boreal weather.

Most of New Hampshire's peatlands are small to medium in size, ranging from less than an acre to about 300 acres. Except in the southeast, the peatlands appear quite similar to those of neighboring Vermont; as in Vermont, no clearly raised peatlands are known.

Locally within New Hampshire the best-known peatlands, typically called "bogs," occupy kettles (or other basins of similar size) and have tree-encircled open mats of plants reflecting fairly nutrient-poor conditions. The mats may be floating, and some sites have central ponds. Interestingly, at one bog (Bottomless Pit near Hanover) at least, a floating mat completely overgrew a central pond within recent memory!

Through much of the state the prominent species are generally the same, although any one site may lack a species or two or have its own particular dominant species. Expect to find various sphagna, including the red *Sphagnum rubellum* and the brown *S. fuscum*, cotton grasses (*Eriophorum* species, especially *E. spissum*), other sedges, heaths such as leatherleaf, sheep laurel, Labrador tea, and bog laurel, the nearly ubiquitous and wiry small cranberry, pitcher plant throughout, round-leaved sundew here and there, and bladderworts in wetter places. Short black spruce, with or without tamarack, may be scattered or even densely clustered on these mats.

Wooded peatlands may have much black spruce, or if somewhat more enriched, tamarack. With increasing elevation or latitude, balsam fir (*Abies balsamea*) occurs, sometimes with spruce or, less commonly, tamarack. Where beaver have been active, alder thickets—perhaps with willows or red maple— may develop on peat. Northern white cedar swamps, whether on peat or inorganic soils rich in organics, have been much logged throughout the state, though a few relatively natural sites survive. At other enriched sites, particularly former bays of ponds and lakes and along sluggish streams, swamps of red maple or even elm (*Ulmus americana*) and black ash (*Fraxinus nigra*) may thrive on peat.

Northern and higher peatlands are clearly boreal in character, not only because of their plants but also because of their animals, especially the birds. One may spot boreal chickadees (*Parus hudsonicus*), black-backed woodpecker (*Picoides arcticus*), gray jays (*Perisoreus canadensis*), or spruce grouse (*Dendragapus canadensis*). As in northeastern Vermont moose (*Alces alces*) may create major trails through peatlands and churn up peats and plants as they graze on favored herbs such as water lilies and browse on willows, alders, and conifers.

To the south and southwest in New Hampshire, species more southern in character grow in, or even dominate, various peatlands. Most evident are the coastal and near-coastal sites now (or formerly, before they were logged) forested with Atlantic white cedar. Other species, generally more abundant in peatlands further south, include poison sumac, the highbush blueberries, Virginia chain fern, maleberry, and skunk cabbage (*Symplocarpus foetidus*). Limited exposures of basic rocks in the state means few richer fens. However, since no comprehensive peatland inventory has yet been undertaken, perhaps significant new sites (or even new types, as in Maine in the last half-decade) will be discovered. There has been modest mining of marl from a few peatlands in the northern part of the state.

Many peatlands in New Hampshire have felt the hand of man, whether in the industrial and agricultural south or the forested lands farther north. Nevertheless, most sites are essentially natural, often passively protected by their relative remoteness. Timbering is the only significant commercial use of the state's peatlands, and this has concentrated on the two cedar species. Some sites with significant natural features are protected by private or public ownership, but efforts to produce a comprehensive statewide inventory and to secure an adequate number of rare and representative peatlands are little beyond their infancy.

MAINE

Maine, although the smallest (32,215 square miles) of the three large northeastern states, has not only the greatest diversity of peatland types but also the greatest number. The diversity is due partly to topography, which ranges from the hilly, island-dotted coast to a mountainous northwest border with Quebec. It is also partly due to glacial geomorphology and soil type (such as marine clays, extensive outwashes, both acidic and basic tills, kames, and eskers). But in large measure it is due to a climate that varies greatly with latitude and distance from the sea, a climate permitting true ombrotrophic raised bogs. Since raised bogs depend on the atmosphere for their nutrients and water, their character is especially sensitive to climatic differences, hence the variety of raised bogs in the state.

Ombrotrophic and nutrient-poor peatlands occur in all parts of Maine, though they are relatively infrequent in the warmer southwest and in the subdued topography of Aroostook County (famous for its potatoes) in the state's northeast corner. Unpatterned raised bogs extend to their southernmost limit in North America in southwestern Maine. Coastal plateau bogs have their southwestern limit on the foggy southern tip of Mount Desert Island in Acadia National Park. Concentrically patterned domed bogs lie in a fairly broad band across the state from about 44° 40′ to 46° 30′, always inland from the coastline by at least a few miles. In the northern half of this band there are also eccentrically patterned domed

Fig. 36 *Maine bog.* This central Maine raised bog displays striking zonation from the bordering lagg through the bog slope shrublands to the noticeably patterned bog plain. Encircling the central pond systems are concentrically arrayed ridges, wet depressions, and some additional ponds. *(Photo courtesy of Maine Critical Areas Program.)*

bogs, some with extensive pond systems. Spectacular when viewed from the air, these sometimes coalesced and usually large (500 or 1,000 to 4,000 acres) patterned bogs occur in adjacent Canada and especially eastward into New Hampshire, where they constitute a significant landform type.

The ombrotrophic peatland vegetation varies greatly in constitution and appearance according to bog type and location on the bog. In places black spruce forms closed canopies. Common are shrub heaths, typically with leatherleaf, sheep laurel, Labrador tea, and a variety of other shrub species. On some bogs lichens (e.g., *Cetraria islandica* and species of *Cladonia*) are prominent, the white species looking like clouds on the mini-mountain hummocks. Finally, on coastal bogs, waving deer's-hair sedge (*Scirpus cespitosus*) spreads like lawns, with sphagnum on the central plain of the plateaus.

The great variety of minerotrophic peatlands have been known best for their general uselessness, difficulty of access, summer bugs, and at certain sites, significant rare plants. Only recently have they been considered systematically from ecosystem and statewide perspectives. Regional distinctions among the state's fens are as yet little studied; the newly discovered, strikingly patterned ribbed fens are not known south of about 45° 30′ latitude (which includes the southern half of Maine, including coastal regions).

Much different in quantity and chemistry, the fens are correspondingly variable in vegetation. Practically every kind of wetland green plant, from mosses and algae to grasses, shrubs, and trees, can be found in a fen somewhere. Laggs bordering raised bogs may have shrublands of rhodora or alder, or they may be wooded with tamaracks. Grassy laggs may have much bluejoint (*Calamagrostis canadensis*), whereas the wettest areas may support beaver ponds and meadows. These same dominants grow on peatlands by lakes and streams, in shallow basins, at seeps, and in other groundwater-influenced environments. Of Maine's treasured rare plants, a number grow in fens; especially noteworthy are plants such as prairie fringed orchis (*Platanthera leucophaea*), northern valerian (*Valeriana uliginosa*), English sundew (*Drosera anglica*), grass-of-Parnassus (*Parnassia glauca*), and bog birch (*Betula pumila*).

In the extreme oceanic climate of the northeastern coast of Maine on the most exposed rocky headlands jutting out into the sea, fogs and winds there spread shrub heathlands with thin, only occasionally saturated, peat soils. In their ability to coat slopes and knolls, these peatlands are reminiscent of the blanket peatlands of western Ireland and Scotland's Hebrides. On the high ridges of a few of the western mountains small peatlands, some with vegetation similar to the raised bogs of lower elevations, fill shallow basins or even lie on gentle slopes. At these high elevation sites, as well as in bogs by the sea in cool and foggy eastern Washington County, ice may remain beneath hummocks well into midsummer.

In a state of such size and such rapidly changing climatic gradients it is not surprising to find geographically restricted species. For example, Atlantic white cedar is at its northern limit in Rockport; at low elevations baked-apple berry (*Rubus chamaemorus*) and black crowberry (*Empetrum nigrum*) come no farther south than midstate; jack pine (*Pinus banksiana*), which grows on a few Maine peatlands, likewise is restricted to the northern coast and higher latitudes as a peatland habitat. As in all of the Northeast's northern states, boreal mammals (such as moose) and birds (e.g., spruce grouse and black-backed woodpecker) are more abundant in or restricted to the northern part of the state.

In the United States perhaps only in the northern Great Lake states has there been as much study of peatlands as in Maine. Prompted by the search for energy and a marketable peatland product, the state of Maine promoted inventories and testing in the early part of this century, again during World War II, and recently following the oil embargoes of the early 1970s. In the 1970s as well, citizen interest in the protection of natural features led to the creation of a state Critical Areas Program and boosted the work of private groups such as The Nature Conservancy and the Audubon Society. Numerous inventories and researches resulted, which, when combined with several conferences, meetings, committees, and program initiations, have led to fundamental baselines of information, discovery of many significant sites, and actual protection of some of Maine's pristine peatlands.

In its peatlands, extraordinary in diversity and frequent throughout the landscape, Maine offers in one reasonably compact place a glimpse of the great boreal peatlands of North America, so prominent from the Canadian maritimes to Alaska. Virtually untouched by man, Maine's peatlands are islands of near-virgin wilderness, unblemished gifts of nature.

This brief summary of the peatlands of the Northeast, only a tiny portion of the globe, is but a hint of what the world holds in store for our discovery. As do the Northeast's bogs and fens, earth's peatlands display immense diversity in form and function and can be found from north to south, from intertidal coasts to mountain heights. Despite their common bond as lands of peat, they are as varied as forests, grasslands, or deserts. The children of climate—molded by topography and matured through millennia—some grow to an independence unmatched by other natural ecosystems.

Truly, peatlands rank high among nature's marvels. Their mysteries are sometimes deep and sometimes distant. In our search for understanding, may our wonder cause our treading feet to leave but the lightest prints.

6 | PLANTS AND ANIMALS: THEIR WAYS TO SURVIVAL

Peatlands are habitats that exclude many species. Of those that do live in peatlands, very few are actually restricted to that landscape. Nevertheless, high water, oxygen depletion, nutrient deficiency, acidity, and unstable substrate are some of the conditions plants and animals must deal with in various peatlands. In this chapter we will look at the special ways some species survive, even thrive, in peatlands.

PLANTS

Plants have evolved many solutions to the problem of unstable substrates. In pools and lakes planktonic algae and submergent aquatics, such as some bladderworts (*Utricularia* species) and naiads (*Najas* species) are simply buoyed by the water. Larger floating-leaved plants, such as water lilies (*Nuphar* and *Nymphaea* species) and pondweeds (*Potamogeton* species), are rooted in the peats with long flexible stems that snake to the pond surface, where large leaves lie like air mattresses atop the water.

Some plants do little more than lie about. *Sphagnum cuspidatum* and some bladderworts grow (not necessarily together) in very shallow pools or on wet peat in some bogs and open fens. Wiry small cranberry (*Vaccinium oxycoccus*) trails among sphagnum and dwarf shrubs. The much-branched pleurocarpous mosses may spread loosely on damp peat (e.g., *Drepanocladus* species) or cover decaying logs as do some species of *Brachythecium*. Almost microscopic, delicately beautiful liverworts (plants related to mosses) thread through sphagna, form thin black skins over moist peat (e.g., *Cladopodiella fluitans*), or creep over woody litter and twigs (e.g., *Diplophyllum albiccus*).

Other tender sphagna and some other mosses grow ever upward, their dying, older parts becoming the newest, uppermost peats. The erect stems bolster one another—sometimes loosely but effectively, as *Sphagnum magellanicum* often does; sometimes, as may be done by *S. fuscum* and *S. rubellum*, forming densely packed hummocks.

Emergent herbs and low shrubs support themselves in a variety of ways, whether in standing water or moist peat. Many grasses and sedges form dense tussocks, a solid core from which the stems and leaves emerge; others are rhizomatous, connected by thick, entangled subterranean stems. Shrubs too, especially those of the heath family, often have profusely spreading, entwined root systems (ask anyone who has tried to cut down through a hummock thick with leatherleaf!). In fact, the root and rhizome jungles in most open peatlands prevent bog-trotters (human, deer, or moose) from sinking through the softer peats. They are so tightly woven that they act rather like the fibers of a trampoline.

Trees growing in peatlands also have some special structural characteristics that seem to help provide them with support. Root systems of species such as red and silver maple (the latter occasionally grows on peats by lakes and rivers) and northern white cedar are shallow and spreading, since they cannot develop below the aerated active zone. Shallow roots provide little support for tall trees in windstorms unless they extend well beyond the spread of the canopy or are well entangled with neighbors in dense groves, a tactic sometimes employed by northern white cedar. Red maple, by progressive dieback of older trunks and multiple stemming from their bases (i.e., shrublike), grow less tall, thus reducing leverage in the brunt of potentially destructive winds. The short stature of black spruce, tamarack, gray birch, and other trees of open peatlands, probably induced by low nutrient availability or high water tables, similarly protects against uprooting and windthrow.

In addition to short stature, black spruce has other support devices. Sometimes the roots become misshapen into "I-beams" that provide strength and reinforcement. The roots also may fuse to make meshes not unlike big snowshoes.

One of the biggest problems faced by plants is surviving in places that are soaked most of the time. With waterlogging there is usually oxygen deficiency, especially where groundwater moves as slowly as it does in most peats. Without oxygen, plant roots cannot respire and thus cannot carry on life processes.

Many sedges solve this problem by producing almost all new roots every year in the aerated upper layer of the mat. Adventitious roots sprout from rhizomes, with maximum growth following high water in the spring. As the water level drops, the roots thread through the aerated peat.

Should water levels rise, several tree and shrub species (e.g., black spruce, northern white cedar, speckled alder, and willows) develop adventitious roots from trunks and stems. Thus, if peats accumulate and water rises significantly during the life of a black spruce, for example, the lower trunk becomes buried ever more deeply, sometimes penetrating several feet into the peat. It has long been observed and accepted that the physiologically active, live roots are those in the upper aerated zone (which of course may be flooded in some seasons). Although this appears to be true for most plants, recent research seems to indicate that at least a few species (e.g., buckbean *Menyanthes trifoliata*) have functioning live roots deep into the anaerobic peats (the catotelm).

Fig. 37 *Layering*. This black spruce (*Picea mariana*) shows rooting of its lower branches, forming a ring of new shoots around the dead parent tree. *(Photo by Charles W. Johnson.)*

Incidentally, black spruce uses adventitious roots not only to solve support and waterlogging problems but also to reproduce and colonize new terrain. Lower branches "layer"— that is, they root amid sphagnum, where they sweep down and lie on the moist vegetation. After rooting has started, other buds become leader shoots, and new erect young trees begin upward growth. Commonly, a ring of young trees forms around the parent, which with time may die, leaving a circle of live spruce. This phenomenon, seen clearly in many raised bogs and very nutrient-poor open peatlands where spruce are subdued in stature, even may repeat itself, forming two, three, or even more (in exceptional cases) concentric rings of trees.

Plants or plant parts submerged in water may develop other structural or physiological modifications to deal with anaerobic conditions. Many plants, including sphagnum mosses, have air sacs and large spaces between cells. Both are places where air can be stored and absorbed by the cells. These air spaces are usually lacking or smaller in the same plants when they are growing in more oxygenated environments or in faster-moving waters where air is mixed better. In addition, some aquatic and semiaquatic peatland plants, such as bladderworts, can take oxygen directly from the water through the entire surface of leaves and stems, not primarily through leaf stomata as terrestrial species do. (The air sacs also may become a source of oxygen for tiny aquatic insects that can pierce the tissue and withdraw air with specialized mouth tubes.)

Many plants can reduce their need for oxygen when submerged, not unlike the strategy of aquatic mammals when underwater. Plants accomplish this reduction through a feedback system, keyed into the concentrations of carbon dioxide (the by-product of respiration). As the level of carbon dioxide increases in its tissues, the plant delivers a signal to reduce the demand for oxygen, thus rationing it over a longer period.

Obtaining and retaining nutrients in surroundings where they are scarce in the first place (notably ombrotrophic bogs) are some of the most difficult physiological challenges to peatland plants. In fact, it is the total inability to cope that keeps most wetland plants from growing in oligotrophic and ombrotrophic sites (and thus reduces the competition for those plants tolerant of low nutrient levels). Some plants meet the nutrient challenges ingeniously, through evolutionary modifications in structure and physiology, through conservation, and through supplementation from outside sources.

In general, nitrogen is available to green plants (and to some fungi, bryophytes, and blue-green algae) in the form of ammonium (NH^+_4) or nitrates (NO^-_3), most of which comes from the activity of bacteria that take nitrogen (N_2) directly

Fig. 38 *Sweet gale (Myrica gale).* A plant sometimes abundant in slightly mineralized peatlands, sweet gale has nitrogen-fixing nodules on its roots. Like its relatives, sweet fern (*Comptonia peregrina*) and bayberry (*Myrica pensylvanica*), sweet gale has a wonderful fragrance. Rub the ripened fruits (end of twig, left) between your fingers to discover it. The male and female flowers (*top and middle insets*) are separate; the latter form seed-bearing catkins (*bottom inset*).

from the air. This process is called "nitrogen fixation." The microorganisms can live independently in the soil or, as they often do, in symbiosis with certain plants (such as sweet gale *Myrica gale* and speckled alder) in the protection of root nodules. However, because of the highly anaerobic and acidic conditions in peatlands, free nitrogen-fixing bacteria are scarce and thus unable to provide much usable nitrogen to the system. What little there is seems to come largely from the breakdown of organic matter and from the wastes of birds and mammals in the area. Traces of ammonium are also available from the air through precipitation.

Nutrient conservation from year to year by peatland perennials is accomplished in many plants either by evergreenness or by nutrient translocation. Spruce, the cedars, and several shrubs retain their leaves for at least an entire year (leatherleaf's previous year's leaves do not fall until new leaves are well established) or for several years (e.g., spruce and cedar). Most liverworts and mosses, including sphagnum species, are essentially evergreen, their leaves and stems remaining intact through winter. Evergreen (perhaps here better thought of as "winter-green") plants have the opportunity to photosynthesize earlier in spring and longer into autumn, thus extending their growing season. Many bryophytes, if exposed and warmed by moderating weather or bright sunshine, apparently may photosynthesize occasionally even in winter.

The translocation of nutrients from nonoverwintering leaves to perennating organs (e.g., roots, rhizomes, corms, bulbs) means that energy-rich compounds such as sugars and starches, as well as essential building materials (e.g., iron, potassium, sulfur) are not lost to litter. Mobilized at the beginning of the growing season, these compounds and elements are available within the plant for rapid expansion of new leaves and in some cases for flowers, stems, or other early developing plant parts. These adaptations have been discovered in many short-season plants, including alpine and tundra species, and are prominent in heaths, sedges, many emergent herbs of richer peatlands, and floating-leaved species rooted in peat-bottomed ponds and marshes (cattails are a well-researched example).

A related physiological adaptation seen in many peatland plants is a reduced need or demand for nutrients to begin with. Some can transfer certain minerals only when they really need them. Phosphorus, perhaps one of the most critical elements to plants' survival in peatlands, is transported to various parts to satisfy demands at various times. Baked-apple berry (*Rubus chamaemorus*), for example, may take up more phosphorus when the growth rate is fastest (early spring) and slow down afterward, then increase use again for flowering and fruiting. They may also transfer phosphorus and other minerals to the seeds much earlier than normal and thereby provide the new generation with a head start.

Another potential problem for peatland plants is the loss of nutrients through "burial" in the deeper peats. The far-spreading and shallow rhizomes and root systems do an effective job of "vacuuming up" nutrients before they are lost forever. The plants also quickly assimilate the substances freed by the activity of decomposer organisms in the active zone.

Because so many bog plants appear to have adaptations similar to desert plants, which withstand real droughts, the German scientist Andreas Schimper theorized in his classic *Physiological Plant Geography* (*Pflanzengeographie auf physiologischer Grundlage*) that bog plants develop adaptive structure to withstand "physiological drought." This theory, introduced in 1898, has remained popular until quite recently. The reasoning is as follows.

The theory proposes that although bogs have an abundance of water, most plants growing there cannot use it because of its acidity. So in effect, bogs are soggy deserts as far as the plants are concerned. To counteract these conditions, the theory continues, the plants evolved special adaptations allowing them to retain water or to decrease their need for it. Some—leatherleaf and bog laurel, for example—have relatively firm and dryish (scleromorphic) leaves, with stomata few and sunken in pits or grooves—adaptations apparently helping to reduce water loss through transpiration. Some species (such as bog laurel) have leaves that curl under at the edges and/or have a coating of fine hairs on the undersides (e.g., Labrador tea), both features also thought to reduce water loss by deflecting or insulating the leaves from drying air currents.

"Physiological drought" seemed a plausible way to explain their anatomical features. Most recent research, however, indicates that physiological drought does not occur in bogs.

Plants show no decrease in their ability to take up water, and the scleromorphic leaves apparently are an adaptation to nutrient deficiency (particularly phosphorus), not water deficiency. The fuzziness of the underside of leaves, on the other hand, is now thought to be a protection against cold, allowing plants to function longer into the fall months or to begin operations earlier in the spring.

Some plants have evolved other ways to deal with nutrient deficiencies, sometimes with the help of other plants and animals—from microbes to larger beings. Many groups of peatland dwellers, including several heaths and most orchids, have symbiotic relationships with fungi, serving as hosts to the fungi and in return receiving nutrients that they could not get on their own. Other groups, well represented in peatlands by sweet gale and alder, have root nodules within which nitrogen-fixing organisms live and change the nitrogen into a form that the plants can use. Yet another group has turned to a completely different source of food—the animal world—by becoming carnivorous. A later chapter discusses these fascinating plants.

ANIMALS

Until recently the fauna of peatlands largely were ignored. Attention focused on the unusualness of the flora and the surface morphologies with which they were associated. We knew that many insects, as well as birds and terrestrial animals, lived in peatlands. We also knew that some species were unusual and that an occasional large and majestic animal might wander through. But with most ecologists concentrating on plants and with the overall infrequency of conspicuous birds and land animals that spend their lives primarily in peatlands, we have remained relatively ignorant of the function or importance of the peatland fauna.

Our assessment of animals has been changing, though, as we begin to view peatlands as complete and complex systems and as we raise our visions beyond seeing them as "useless," "peculiar," or "quaint." Much new information has come from recent research, a great deal of it stimulated by proposals for extensive fuel peat mining.

The impression of the Northeast's peatlands (especially bogs and poorer fens) as being rather barren of wildlife is largely accurate though somewhat exaggerated. Relative to most other ecosystems, nutrient-poor peatlands offer fewer rewards to foragers or predators. Likewise, the living conditions are rather inhospitable to many smaller creatures who, by virtue of their size, must live almost entirely within their bounds. In fact, some of the same problems confronting plants in peatlands also confront animals: extended periods of high water or saturation, a dearth of nutrients, lack of oxygen (for fossorial or burrowing animals), and acidity (particularly in bogs).

Winged animals and terrestrial animals with great mobility have a fairly simple solution to peatland limitations: they find what needs can be fulfilled in peatlands, then move on to other places. Large, roaming herbivores such as moose (*Alces alces*) and white-tailed deer (*Odocoileus virginianus*) often may enter peatlands for browsing and grazing, sometimes making prominent trail systems by their regular travels. However, the peatlands apparently provide only a fraction of the animals' food and seldom are used for lounging or breeding; these cervids move on to other landscapes to satisfy their life's needs. Far-ranging predators such as bobcats (*Felix rufus*), fishers (*Martes pennanti*), and coyotes (*Canis latrans*) may traverse or occasionally make an extended visit to peatlands in their search for small mammal prey. But for them too peatlands are incomplete habitats. The obvious mobility of birds provides many species the opportunity to seek food or nesting sites beyond the limitations many peatlands impose. When small mammals are abundant, raptors (both hawks and owls) may spend much time in peatlands, but it is easy for them to

Fig. 39 *Lincoln's sparrow (Melospiza lincolnii)*. Rare in most places in the Northeast, this streaked sparrow is relatively common in the large northern bogs, where it frequents the "dwarfed" forests and shrubby thickets. A ground nester, Lincoln's sparrow is almost always found near water—in swamps, at pond edges, or around and in bogs. Its song may be confused with that of the marsh wren (*Cistothorus palustris*), with which it often occurs.

fly to other terrain at any time. Even small nesting species such as Savannah sparrows (*Passerculus sandwichensis*) and palm warbler (*Dendroica palmarum*) are free to fly to outlying areas to find food, then to return to their peatland nesting site. Of course, the vast majority of birds that use peatlands in the Northeast are migrants, remaining for only part of any year.

Numerous small animals, both invertebrates and vertebrates, are in effect "rooted" (for part or all of their lives) in a fairly small area of peatland and do not or cannot leave when conditions get difficult.

For air-breathers lack of oxygen in their peatland habitat is a crucial problem. The three-dimensional home of nonflying forms may be compressed into the upper few inches of mat (the acrotelm), the surface, and erect plant parts, some or all of which may be flooded during part of the year. The lower waterlogged peats, starved of oxygen, contain only a few microscopic organisms tolerant of anaerobic conditions. To survive where water levels rise and fall, animals must be able to extract the limited oxygen from the water, get above the water, conserve oxygen, or do without oxygen for extended periods. Indeed, during exceptional high-water episodes, many small animals such as ants and voles drown.

Many aquatic organisms in peatland waters carry their own air supply with them as they skim over the surface or dive in search of prey, plants to eat, or detritus to scavenge. Whirligig beetles (*Dineutes* species), for example, which are common in many still waters or quiet sections of lakes and ponds, trap a bubble of air under their body plates and take it down with them when they dive, using it as a human scuba diver uses an oxygen tank. The water spider (*Dolomedes urinator*), seen regularly skating over the water surface of minerotrophic peatlands and other wetlands, has an air supply trapped in body hairs when submerged on its predatory hunts. Pond snails (order Pulmonata), some of which are common in fens, spend a great deal of time underwater scraping for food. They store air under the shell and, when low on oxygen, drift to the surface and replenish their supply through specially designed, extensible "periscopes."

Most small warm-blooded animals of peatlands avoid the anaerobic depths altogether. Shrews, moles, voles, lemmings, and some birds that spend at least some time within the peats or tangles of roots, rhizomes, and decaying vegetation, confine most of their hunting, grazing, food-gathering, or nesting to the upper, somewhat drier levels. Consequently, in open peatlands runways often can be found along the sides of hummocks, and nests can be discovered within or atop hummocks. Where there are isolated or abundant woody plants, their fallen limbs and trunks often provide good small-

Fig. 40 *Water shrew* (*Sorex palustris*). Wooded peatlands with open water are among the many wetland haunts where this large insectivore runs or swims in search of prey. Its large and partially webbed hind feet enhance its great dexterity in water. Observers have reported that it can run across the surface of water for some distance, buoyed by those feet.

mammal habitats above high-water limits. Many resident nonaquatic insects also remain at, near, or above the water surface, feeding in the acrotelm or among the live plants. Overwintering insects may be adults, larvae, pupae, or eggs in above-ground plants or plant parts, such as in the leaves of pitcher plants, in the seed heads of sedges, or under the bark of trees.

Acidity apparently is a major cause of the depauperate fauna of bogs, since it can be deleterious or lethal to the adults, eggs, or immature stages of many groups of animals. Certain life stages in many fish, amphibians, and soft-bodied invertebrates (such as earthworms, mayfly larvae, and mites) are very sensitive to acidic waters: eggs may become malformed or sterile; the metabolisms and reproductive capabilities of adults may be impaired; and in fish and amphibian larvae, gills may be injured to the point where they can no longer function. Moreover, the brown humic acids in bog water can actually precipitate on body parts, clogging gills or other vital organs.

The animals that spend some part of their lives in these acidic conditions have developed physiological resistance to acidity. Hard-shelled or thick-skinned invertebrates such as beetles (family Carabidae) and centipedes (class Chilopoda) are the most abundant animals seen scurrying around the mat. Some kinds of earthworms are apparently somewhat immune to the effects of the acids. The few amphibians that live in bogs or at least lay eggs there, such as the four-toed salamander (*Hemidactylium scutatum*) and the wood frog (*Rana sylvatica*), are tolerant of low pHs, even down to 4.0.

The ecological implications of changed food chains and webs are great. Where normally abundant "lower" organisms are scarce or absent in peatlands, higher animals that depend on them as food cannot survive either. In such situations other organisms, such as bacteria and fungi, may take the functional place of those that might otherwise be there.

So peatlands do not just modify—sometimes greatly—the landscape in which they develop; they often create demanding habitats for their inhabitants. In the next few chapters we shall look more closely at some of the groups of plants and animals that hold our fascination in peatlands: some because of their abundance, some for their adaptations or rarity, and still others because of the intangible but certain joy felt with each encounter.

7 | SPHAGNUM MOSSES

Sphagnum is a large and important genus of mosses the species of which occur worldwide. These so-called peat mosses are the most abundant plants in many northeastern peatlands, particularly raised bogs and other very nutrient-poor sites. Throughout the region an avid and knowledgeable searcher could discover more than fifty species, but during a day's visit to a particular peatland, the average visitor might encounter no more than five to seven species. Some species are identifiable by color, size, denseness, or habitat (hummock or hollow, edge or center of peatland, and so forth).

Color is the most noticeable feature in distinguishing types, but it is by no means totally reliable, since hues can vary within a species (often according to degree of shading or season of the year). Several sphagna have breathtakingly beautiful colors, especially when glistening under a soft spring rain or glowing in the warm shades of a twilight sunset. None is more striking than *S. rubellum*, which can carpet large areas with brilliant burgundy reds. Some of Maine's and northern New York's larger bogs are resplendent with the reds of *S. rubellum* and the closely related *S. nemoreum*.

Species of sphagnum are segregated by a combination of

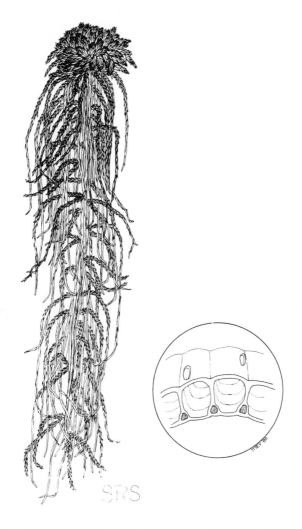

Fig. 41 *Sphagnum moss*. Along the stem of sphagnum plants grow individual leaves and fascicles of leafy branches. In this representative species there are three branches per fascicle. The head (or capitulum) contains young branches, sexual organs, and the perennating terminal bud. The enlarged cross section of a leaf (*inset*) shows the large water storage cells and the small, triangular (*in section*) photosynthetic cells. Sphagnum species differ in compactness, size and shape of the capitula, orientation and number of side branches, color, and other anatomical features. *(Adapted from Andrus 1980. By permission of New York State Museum.)*

microscopic and macroscopic characters. Some species can be quickly learned just by means of a few macroscopic features. Only *S. squarrosum* (a species of shaded, wet places rich with decaying leaves and twigs), for example, regularly has leaves at right angles to the branches. On the other hand, many species require considerable expertise and microscopic inspection to ascertain identity.

Sphagnum plants have a fascinating growth habit: they grow from the top and die at the bottom. The top (capitulum) of a single plant typically grows above water level, while below, submerged in water or removed from sunlight, the plant is dead and beginning to decay, directly contributing to the peat deposition. The capitulum of sphagnum contains the terminal bud, new branches, and at times, sexual organs. The main stem is several cells thick. Scattered along it are small stem leaves and fascicles (groups) of leaf-covered branches. Some branches are pendant and lie along the stem, others spread outward. The branches are a very few cells across, and the leaves are essentially one cell thick. Leaf cells come in two completely different types: large empty ones that store water and very much smaller chlorophyllous ones that produce food for the plant and give it its color.

Spores develop in dark spheroid capsules on short pale stalks atop the plants. Though many sphagnum species can produce spores for reproduction, few are known to do so regularly. It is believed that reproduction occurs largely by vegetative dispersal, stems or branches being transported long distances (perhaps by ducks, large mammals, or wind). Nevertheless, a number of species freely produce spores. A careful observer in summer may be treated to the occasional sound of popping spore capsules as the lids fly off!

In the leaves and stems of sphagnum are large specialized cells that at maturity become empty and spacious. Their cell walls may have thickening in the shape of spirals, rings, or plates. These structures provide strength and some rigidity to the rather simply constructed plants. The large cells, without cytoplasm, are miniature buckets, capable of holding large quantities of water.

Picking up and retaining water is an exceptional ability of sphagnum. Capillary pathways are formed by the leaves along the branches, between the branches, and particularly along pendant branches that lie beside the main stem. Since water transfer is a mechanical process, both the lower dead parts and the upper living portions act like wicks.

Some species of sphagnum can hold vast amounts of water, as much as twenty-five times their own weight! This attribute, along with others to be discussed shortly, makes sphagnum an important agricultural and horticultural aid, helping to keep soils moist. This absorptive ability explains why Native Americans used dried sphagnum as diapers. It was also used by the Allies in World War I and by the Japanese in their war with Russia as a field surgical dressing for wounds. Dried sphagnum proved more than twice as efficient as cotton in absorbing fluids; moreover, its acidic character and bog origin rendered it naturally sterile!

Another important characteristic of sphagnum is its ability to hold onto ions (potential nutrients) with which it comes into contact. Sphagnum leaves can pick up calcium, sodium, or other mineral ions, secure them tightly, and give up hydrogen ions in return. Hydrogen ions, set free in water, contribute to the acidity of the peatland. These bound ions are buried with the dead sphagnum as it becomes peat. Those not released during partial decomposition are thus entrapped in the peatland. If peats are taken from their natural setting and allowed to rot, the ions are liberated. (Peats formed from other plants, of course, share this property to varying degrees, depending on the extent of decomposition and the original chemical structure of the plants themselves.) Thus, sphagnum is a superlative soil additive—it is a kind of timed-release fertilizer, wrapped in ready-made mulch and soil conditioner.

Sphagna vary tremendously in rates of growth, depending on species, site, and climate. The slower-growing species may elongate less than a centimeter a year; faster-growing species (notably some aquatic species) under good conditions can add as many as fifteen to twenty centimeters a year.

Species presence and growth rate depend on site factors such as shading, acidity, available minerals, drought, wetness, and flooding regime. Each species has its own need thresholds and tolerances for each condition. Since some species have well-known requirements, they are excellent indicators of peatland character (for example, the degree of minero- or ombrotrophy) and local climate (such as the degree of maritime influence).

Research has shown that many sphagnum species occur only within specific ranges along pH, nutrient, and wetness gradients, which sometimes correspond with successional sequences. Among these gradients the relationship between sphagna and wetness is crucial and complex. The ecological significance of this relationship is probably profound, of importance not only to the sphagna but also to the formation of the living mat and hence to the character and very survival of many peatlands. This is nowhere more true than in sphagnum-dominated raised bogs.

Expressed in oversimplified form, this relationship rests on the fact that some species of sphagnum prefer water more than others. Some favor the low, moist, often-saturated hollows; others grow up and out of direct contact with water; still others span the intermediate distances between the low and high places.

In the wet hollows of the treeless portion of a peatland two species may occur. *S. recurvum* grows in damp hollows. The leaves on the branches are arranged in five neat rows. This species ranges in color from green to light brown and often appears nearly white in hollows that have dried out. A second species that grows in the wettest portions of the peatland is *S. cuspidatum*, always green, whose branches look like small wet paintbrushes because the long branch leaves are all bent in the same direction.

Five species commonly form the open mat and hummocks of a peatland. *S. magellanicum* is in the wettest portion of the peatland; it is dark red and has long, tongue-shaped stem leaves, the largest of the open-mat species. *S. centrale* is similar in appearance but a lighter red, with smaller leaves. An even smaller red species, *S. rubellum*, forms mats in places that are never under water. This species may be recognized by its color, its small five-rayed capitulum, and by branches that have leaves arranged in five rows. Still another red species, *S. nemoreum*, grows on hummocks in drier places of the peatland. (*S. nemoreum* is distinguished from *S. rubellum* by its pom-pom-shaped capitulum and branch leaves not arranged in five rows on the branches.) *S. fuscum* grows in the driest of all the habitats in the peatland. This small species is the only truly brown *Sphagnum* and often forms large, dry, tightly compacted hummocks.

The prostrate, open-branched, limp sphagna of hollows are designed to maintain the greatest possible contact with water. In contrast, the taut and upright hummock sphagna support themselves vertically; they protect themselves from overdrying by being densely packed, exposing a minimum of surface to the sun, air, and wind.

When dry periods come, as they inevitably do, hummock and hollow species react differently to the stress. The tight-packed hummock plants, with their large and highly absorbent pore spaces, can retain moisture longer than the fine-pored and exposed plants in the hollows. Thus, the hollows dry out faster than the hummocks. But the hollow species are adapted to this condition: they can withstand several days of desiccation, and revive and grow quickly on being watered. The hummock species, in contrast, can withstand drying out better than the hollow species, but once dried they are not so easily revived. So as long as water is available, the peatland can continue to build, but if water is in short supply, the peatland assumes a defensive strategy. The hollow species stop growing but are ready to begin again immediately should conditions improve; the hummock sphagna hold steady for as long as they can but will retrench and even die if the dry spell is protracted. Ultimately, this constantly shifting accommodation between plants and water expresses itself in a peatland's expansion, maintenance of a steady state, shrinkage, or even death.

So the role of sphagna in peatland is large and various—as highly important builders of age-old peat, as dynamic indicators of environmental conditions within and without, as wondrous sculptures of innumerable sizes, shapes, textures, and colors that help make peatlands so subtly beautiful to us who look upon them somewhat as art of a grand nature.

8 | THE CARNIVOROUS PLANTS

No group of plants has enthralled us more than the carnivorous species—those that have turned the tables on the animals, eating them rather than being eaten. From the mythical man-eating tree of Madagascar to the widely known Venus's-flytrap of eastern North America, carnivorous plants hold a special place in the botanical world, and in our fancy. The "civilized" world has been fascinated by them for more than 2,000 years, and peatlands are home to many.

Better than 450 species of carnivorous plants are found worldwide, many of which are in North America, and many of those thriving in wet or impoverished environments. But there is little similarity in distribution of the various genera and species. Sundews (*Drosera* species), butterworts (*Pinguicula* species), and bladderworts (*Utricularia* species), for example, are widespread throughout the world, on many continents. Others, like the pitcher plants (*Sarracenia* species) of North America, are restricted to a single continent. Still others have even more limited ranges. An excellent example is the Venus's-flytrap (*Dionaea muscipula*), which, before its attractiveness as a novelty led to sales in stores everywhere, grew only in the Cape Fear region of North Carolina.

In the Northeast, more than 20 species of carnivorous plants may occur, not all in peatlands, however. In general, the number of species increases southward and/or toward the Atlantic coast. In North America the greatest diversity is in the southeastern Coastal Plain, from eastern Virginia to eastern Texas.

As mentioned in Chapter 6, plants may have evolved carnivorism as a way to obtain nutrients in places deficient in nutrients in more normal sources. Besides the essential protein-builders phosphorus and nitrogen, the plants may well get vitamins and other trace minerals from digesting insects and other animals. For the most part, carnivory does not appear essential for the survival of individual plants—recall that they all photosynthesize—but rather helps them to remain vigorous, grow larger, produce more flowers, and be more fertile.

The use of the word *carnivorous* instead of *insectivorous* is intentional, for the plants consume more than insects. In addition to flying, walking, and crawling insects, the world's carnivorous plants count isopods, mites, spiders, small fish, and even frogs among their victims—though, of course, no one species is so catholic in its diet.

There have evolved active and passive ways to trap prey. Passive methods include pitfalls (such as the leaves of pitcher plants)—the animal simply falls into a trap that does not move or respond—and "flypaper" where sticky substances on the leaves (sundew leaves have such substances on the ends of hairs) hold victims. Active traps involve some kind of immediate action by the plant to ensnare the victim. The jawlike trap of the Venus's-flytrap is the most renowned of these; its leaf literally imprisons prey as it snaps shut, spinelike protrusions acting as bars. Another active trap is that of the bladderworts. Their submerged tiny sacs or bladders can inflate rapidly, sucking in passing aquatic organisms.

In order to use these wonderfully ingenious traps, carnivorous plants attract prey by deception or by promise of reward. Some leaves, like many flowers, have nectar glands that

So subtle and arresting are the autumn colors of many
peatlands that the landscape becomes a tapestry of splendor.
(Photo by Richard Czaplinski.)

The seed heads of cottongrass
(*Eriophorum species*) seem almost suspended above the mat of
a Massachusetts peatland. (Photo
by Marc DesMeules.)

Rose pogonia (*Pogonia ophioglossoides*) is a companion of a few
other orchids in semiacidic open
peatlands. (Photo by Stephan Syz.)

An October flyover reveals the distinct outline of a peatland in west-central Vermont. The perimeter is sharp with the turning color of red maple (*Acer rubrum*), but the open mat of shrubs and sphagnum mosses remains brown and green. (Photo by Ian Worley.)

An open-grown pitcher plant (*Sarracenia purpurea*) displays its typical red leaves, awaiting its next insect meal. (Photo by Richard Czaplinski.)

The yellowing needles of tamarack (*Larix laricina*) are encased in a fall morning frost at a Vermont peatland. (Photo by Charles W. Johnson.)

Leatherleaf (*Chamaedaphne calyculata*) in the foreground, pondweeds (*Potamogeton* species) in the middle, and black spruce (*Picea mariana*) and tamarack (*Larix laricina*) in the background frame this serene kettlehole bog in northern New England. (Photo by Charles W. Johnson.)

Fragrant water lily (*Nymphaea odorata*) is a common floating-leaved plant of peatland waters. Its leaf bases are a delicacy to moose. (Photo by Stephan Syz.)

An unfortunate damselfly is destined to be a meal for this spatulate-leaved sundew (*Drosera intermedia*), as the leaves curl around it. (Photo by Stephan Syz.)

The early-blooming rhodora (*Rhododendron canadense*) graces many northeastern peatlands and boreal uplands. (Photo by Stephan Syz.)

The blossom of yellow lady's slipper (*Cypripedium calceolus*) almost glows in the deep shade of a Vermont northern white cedar swamp. (Photo by Susan Antenen.)

Arethusa (*Arethusa bulbosa*) is a rare peatland orchid in the Northeast; its small but exquisite June blossom hides in semi-open poor fens. (Photo by Susan Antenen.)

entice insects. Pitcher plant leaves emit a fragrant scent; those of butterwort issue a musty, funguslike odor that appeals to different prey, such as small flies. The bright venation patterns of pitcher plant leaves apparently lure potential quarry, including bees and ants, into the inescapable depths, a visual path reinforced by a trail of nectaries. Sundew's glistening droplets of sticky mucilage may well be visually attractive (though ultimately fatal) to curious insects, such as damselflies and many others.

In the remainder of this chapter let us take a closer look at species of the three most frequently encountered genera in the Northeast—pitcher plant, sundews, and bladderworts.

PITCHER PLANT

The family of pitcher plants (Sarraceniaceae) has the largest carnivorous plants—the leaves of one southeastern coast species, *Sarracenia flava*, may be almost three feet tall. This family in North America contains nine species: seven are in the Southeast, one in the Pacific Northwest, and only one north of Virginia. That species, *Sarracenia purpurea*, found throughout the Northeast, is called simply "pitcher plant."

Pitcher plant grows from Florida to Labrador (except for a gap in Georgia) and from the Atlantic seaboard to Louisiana, Minnesota, and Saskatchewan. It has two distinct subspecies: the northern-distributed *S. purpurea purpurea* and the more southern *S. p. venosa*. Both subspecies flourish in the Pine Barrens of New Jersey and often can be found together elsewhere in the Northeast.

Until the 1800s scientists were uncertain about the function of the family's vase-shaped pitchers, which usually contained some water. Linnaeus speculated that they were water-storage vessels that sustained the plants in droughts. Others thought the large hood folded down over the mouth to create a temporary hiding place for insects as they sought to elude predators. Now we know that the leaf is highly evolved and

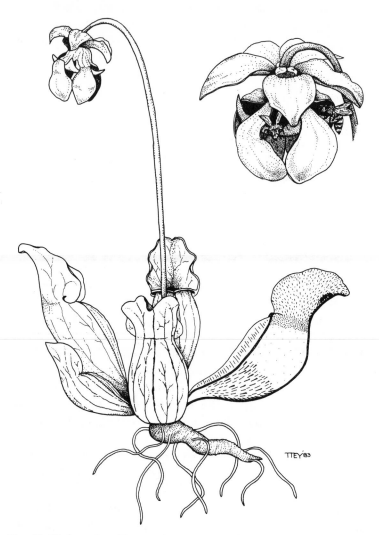

Fig. 42 *Pitcher plant (Sarracenia purpurea)*. The intricate flower structure ensures cross-pollination. The pollinator (typically a bumblebee) enters between the petals by one of the apices of an umbrellalike stigma; as it passes the stigma, it brushes off pollen gathered from another pitcher plant flower; the bee exits by a different route (at the base of the petal, below the stigma) so that pollen gathered inside does not come into contact with the stigma. The cutaway at the right shows the five zones of the leaf.

specialized, with edges that have curled around and fused to form a liquid-holding vessel. *S. purpurea* leaves grow from a basal rosette, all originating from an upward-arching central "core." A "keel" provides structural reinforcement to each leaf so that the opening is always upright. Other species of the family have solitary leaves that stand erect like columns.

The pitcher plant leaf is marvelously designed to entice and then entrap its victims. It has five distinct zones, ranging from top to bottom, each with a discrete function. The topmost zone is a flared flap, spread to welcome "guests." Nectar glands and bright venation attract the insects, while hundreds of downward-pointing hairs on the flap (described as miniature punji-sticks) encourage descent into the pitcher and make climbing out difficult or impossible.

The second zone is a rather smooth rim or "neck" inside the pitcher at the constriction below the flap; this is made of epidermal cells with thickened walls, laid on each other like shingles on a roof. Because the cells are both adhesive and easily dislodged, they stick to the feet of many prey species. Thus, when an animal descends from the flap and reaches this zone, it begins to slide on the smooth vertical sides. In scrambling to get back up, it accumulates globs of sticky cells on its feet, further hindering escape. Even a flying insect may have trouble taking off, especially when trying to get through the rather narrow opening of the rim. Eventually, often out of pure exhaustion, the prey slides down into the third zone, the smooth main body of the pitcher. (Rainwater caught by the leaf usually fills this zone.) There the prey dies, often by drowning. The smooth walls of this zone contain glands that secrete digestive enzymes during portions of the year. Thus, entrapment is completed and digestion begun here.

In the fourth zone, at the bottom of the pitcher, the prey finally comes to rest, is restrained by even more hairs and further digested by additional glands, and may be absorbed by the pitcher plant. The assimilation of animal material into plant tissue is speeded here by thinner cell walls (except around the bases of the hairs).

The fifth and lowest zone of the leaf is the long, narrow stalk. It is without glands and serves as a receptacle for indigestible animal parts, mostly the chitinous exoskeletons of insects. Vascular tissue in it conducts some of the products of digestion and photosynthesis to the roots and particularly to developing flowers and seeds.

With pitcher plant the need for animal supplements seems to be associated primarily with flower and fruit development (hence reproduction), since the plants reduce or even halt secretion of digestive enzymes after seed production. During periods when it is digesting, the plants inject more, and more acidic, digestive fluids into the water held by the leaf whenever there is dilution by rain or addition of more prey. Animals trapped in water in the leaf of S. *purpurea* are digested not only by enzymes produced by the plant but also by bacteria living in the water. Other species of pitcher plants, such as the cobra plant (*Darlingtonia californica*) of the West Coast, lack the enzymatic function; bacteria do all the digestive work.

Ironically, some insects and other small organisms survive perfectly well inside the pitcher. They may spend part of their life there and move on or, in a few cases, make it their permanent home. A common example is the small mosquito, *Wyeomia smithii*, the larva ("wriggler") of which lives in the very fluid that consumes other kinds of insects. It may even hibernate through winter, secure in the leaf-trapped ice. Another insect, a fly of the genus *Sarcophaga*, as a larva feeds on the remains of insects that have died in the pitchers. In still another instance, a species of small wasp builds its nest right inside the pitcher. These organisms may have antienzymes to counteract the digestive enzymes in the fluid. Aquatic forms may be less susceptible to the action of the enzymes, at least as long as they are alive. Most trapped and digested animals are terrestrial species, initially debilitated by wetting agents produced by the plant. They simply drown, and once dead they readily decompose.

Pitcher plants provide other wonderful illustrations of how

plants and animals have evolved together in curious and unexpected ways. Several predators use the pitcher plant as a
place to ambush their prey, taking advantage of the plant's attracting colors and scents. Certain spiders, for example, spin
webs across the mouths of the pitchers to catch entering insects. Even small tree frogs (*Hyla* species) have been known
to sit on the leaf rim, waiting patiently to capture insects as
they pass by; sometimes the tree frogs fall into the pitchers
and become victims themselves! Perhaps the most fascinating
story is that of small nondescript moths of the genus *Exyria*,
but that will be a tale to be told later, in Chapter 11.

THE SUNDEWS

Sundews, bearers of tiny spine-borne diamond droplets,
glittering in the first sun shafts of a windless dawn, have long
been my favorites. So small yet so wondrously captivating, to
me they seem more than just plants—they are the jeweled
sprites in miniature worlds of dwarf shrubs and mosses.

Sundews belong to the family Droseraceae and live in many
peatland habitats of the world. About one hundred species
are known, with the greatest number in Australia and southern Africa. The Northeast has five species, only two of which
are widespread and fairly common. Round-leaved sundew
(*Drosera rotundifolia*) is the most frequently encountered.
Acidic and treeless peatland is a favorite habitat, but it also
does well in many open and rather barren areas, such as sandy
road embankments. The other common species, spatulate-
leaved sundew (*D. intermedia*), is more restricted to peatlands, especially those near the coast. It often grows in the
same peatlands as round-leaved sundew but in somewhat
wetter and slightly more minerotrophic locations.

The other three species are more restricted in distribution
and are rarer in the Northeast. Linear-leaved sundew (*D. linearis*) occurs mostly in minerotrophic peatlands from northern Maine to the upper Midwest and in Canada. Also north-

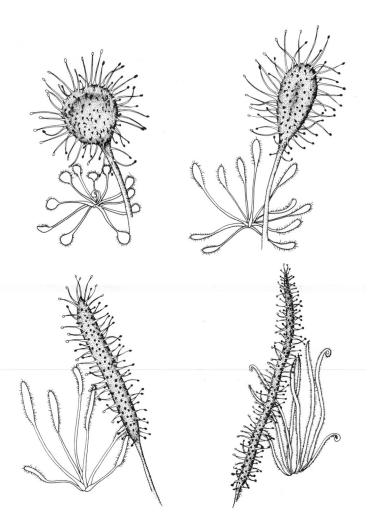

Fig. 43 *Sundews (Drosera* spp.). Illustrated are four of the five sundew species that grow in the Northeast. *Clockwise from top left*: round-leaved (*Drosera rotundifolia*), spatulate-leaved (*D. intermedia*), thread-leaved (*D. filiformis*), and slender-leaved (*D. linearis*). Emanating from a basal rosette, sundew leaves lie close to the ground or arch upward. The shape of the leaf blades helps distinguish the species (the leaves are drawn to show shape and arrangement but are of different scales). Unlike the others, thread-leaved sundew grows mostly in habitats other than peatlands down the Atlantic Coastal Plain and has a deep purple blossom (instead of white).

ern is the English sundew (*D. anglica*), whose range barely extends from Canada into Maine. Thread-leaved sundew (*D. filiformis*) is a Coastal Plain species, ranging from southeastern Massachusetts to southern New Jersey. Primarily an inhabitant of wet sands, it sometimes grows on sandy peats.

Our sundews are believed to be of southern (near-tropical) origin. This may account for their slow growth in northern peatlands during the cooler seasons of spring and fall, when sphagnum mosses with which they grow do very well. But in the warmth of summer sundews grow very rapidly. Thus, in the cool months sphagna often overtop the sundews, but in midsummer the sundews gain the upper hand. In such circumstances a living rosette of sundew leaves sits atop a series of dead rosettes along a long stem (perhaps with some intervening strands of sphagnum), indicating the back-and-forth cycle of growth and suppression. The series of rosettes also may be the result of sundew's sensitivity to frost. Any ground frost is usually enough to kill the leaves (thus, in post-frost autumn we find sundews by looking for the erect seed-bearing stalk). The plants overwinter via their live roots and hibernating leaf buds that emerge again the following spring.

Like most carnivorous plants, individual sundews do not require a diet of animal protein to survive or reproduce. However, carnivory appears to enhance both activities. If no insects are being consumed, sundews can bypass seed making, reproducing asexually by tubers, axillary buds, or leaf buds.

Sundews snare their prey passively, by means of viscid "tentacles" (actually glandular hairs) to which insects become stuck. Once an animal is trapped by the longer, outer tentacles, the plants respond actively by folding those tentacles (and in some species the entire leaf) around the bodies. The slow tentacle "movement" is in reality a phenomenon of fast growing; the tentacles add cells and are longer after the folding than before it. The overall action seems to be a way of increasing the leaf's surface contact with the insect, thereby increasing the rate of digestion and absorption.

The speed of tentacle folding is highly variable, depending on what is being devoured. If the victim is an active midge or small bug, the tentacles can respond rapidly, as soon as 1½ minutes after the first stimulus, and entirely wrap the animal in less than twenty minutes. But if the substance touching the tentacles is organic matter that does not move (such as a dead insect, a piece of meat, or a grain of sugar), the response is much more sluggish, taking several hours or days to complete. The movement of the victims seems to affect the rate of closure; dead flies, for example, do not provoke the same response as struggling live ones. If inanimate objects (such as grains of sand) are dropped on the leaves, the tentacles may not move at all, or they may begin to fold but stop and open again, as if realizing they've been fooled.

Finally, after digestion is complete, the leaves unfurl and the tentacles fold back. The trap is set for the next meal. However, a leaf can repeat this sequence only a few times during its lifetime.

Shorter hairs on the inner surface of the leaves secrete a mucilage that contains the digestive enzymes and apparently an anesthetic to debilitate prey. (Charles Darwin once noted that the droplets not only attracted insects but seemed to stupefy them.) As the captured animal becomes digested into soluble materials, some (certainly at least protein- and mineral-rich molecules) are absorbed into the leaf cells. From there they are distributed to other parts of the plant. At times, especially when the leaves are packed with lots of insects, bacterial digestion assists in the breaking down of the animal body.

THE BLADDERWORTS

The bladderworts (genus *Utricularia* of the family Lentibulariaceae) have perhaps the most sophisticated scheme for capturing prey. This inglorious-sounding genus is one of the most astonishing yet least noticed groups of carnivorous plants.

Bladderworts are found the world over, from the arctic to the tropics, and in all the continents (except Antarctica). The 14 species known in the Northeast vary in structure, habitat, and distribution. They may be free-floating in still or slow-moving waters (e.g., common bladderwort, *U. vulgaris*); they may be semiaquatic in muds or peats (e.g., horned bladderwort, *U. cornuta*); or they may be even epiphytic, supporting themselves on or within wet mosses (e.g., *U. subulata*).

Two species are quite widespread and common in the Northeast: horned bladderwort and common bladderwort. The former, named for a horn or spur that hangs down from its flower, grows in bogs and fens alike, in moist-to-wet sites, in sands or peats, and even in some open, still waters. The latter is as wide ranging and nearly as diverse in habitat, though it tends to be more aquatic. Its flower lacks the spur, and its bladders are larger than those of many other bladderworts.

Several species are much more restricted in range or habitat within the Northeast. Lesser bladderwort (*U. minor*), for example, though found more or less throughout the region, is nowhere common. It tends to favor hollows and saturated flats of minerotrophic peatlands. The totally aquatic purple bladderwort (*U. purpurea*), frequent to the north in adjacent Canada, becomes scarce southward, generally confined to the Atlantic Coastal Plain. In the New Jersey Pine Barrens it is sufficiently scarce to be listed as an endangered species.

All bladderworts are entirely without roots. The stem and finely divided leaves provide support for the stalked, aerial, and often handsome flower. Except when flowering, the plants usually are not noticed by most wetland visitors because of their subsurface obscurity.

The plants owe their name to tiny sacs (utricles) on their leaves. These bladders once were thought to be "flotation devices" and a way for the plants to get air in an aquatic environment. But actually they are miniature, exquisitely designed traps (0.3 to 5 millimeters in diameter, depending on the species) for aquatic organisms such as isopods, water fleas, mosquito larvae, or anything else that might fit, includ-

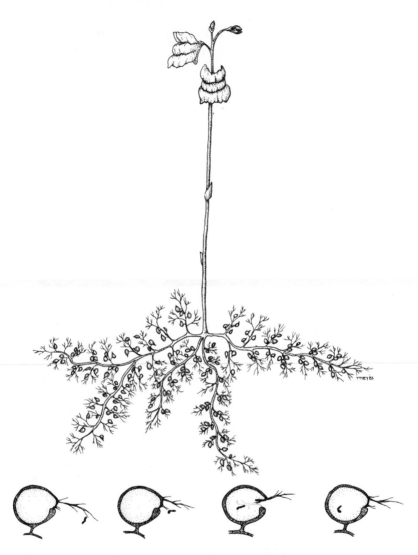

Fig. 44 *Common (or greater) bladderwort* (*Utricularia vulgaris*). This bladderwort occurs throughout the Northeast, mostly in aquatic or semiaquatic habitats. The yellow flower, looking somewhat like a snapdragon's, rises above the submerged and much-branched stems. The plant has no roots, obtaining its nutrients directly from the water or from prey captured in the small bladders. The lower drawings depict the capture of a small aquatic animal, which, having brushed one of the triggering hairs, is sucked into the bladder where it will be digested by the plant.

ing (in the case of the larger bladderwort species) very small fish and tadpoles.

The bladders have been compared to little "stomachs" because they hold and digest food. Yet they are more than stomachs, since they do not simply receive food; they capture it. Years of research and investigation have revealed much about how the traps work, but many questions still remain about the exact details of the operation.

Although the bladders may differ in their size, shape, and specific mechanics, the basic principle is the same in all species. The bladder is a chamber into which an organism is sucked forcibly and ultimately digested. Each sac has a small orifice through which the organism is drawn. When the sac is not operating, the orifice is closed off by a downward-hanging "door" and a smaller upward-directed flap (velum). In one group of bladderworts (including common bladderwort) small hairs project above the door; in the other group the hairs are lacking (as in the horned bladderwort).

The capture of prey appears to involve a purely mechanical series of steps, each accomplished with great rapidity. At rest, the bladder is not fully rounded but indented on both sides. The door is shut tightly against the threshold, with the flap sealing the bottom by a sticky secretion. The hairs project outward. Potential prey swim toward the bladder orifice, perhaps attracted by sugar issuing from special glands on the outside of the door. If an animal touches one of the trigger hairs while near the opening, the bladder springs into action. The hairs attached to the free edge of the door, when twisted slightly, cause the seal to be broken between the orifice edge and the door—much the way a vacuum seal is broken by cracking a jar lid a bit. (Another theory contends that touching the hairs causes an electrical stimulus to be passed to the bladder.) When the seal is broken, the water rushes in past the inward-swinging door, and the current carries the victim with it. Sometimes it is possible actually to hear this action by lifting a plant out of the water and holding it close to your ear. Listen for a faint popping as sacs gulp air instead of water.

As the sac fills with water (and victim) and the pressure is equalized inside and out, the door automatically closes by a kind of spring mechanism. The creature is caged, ready to eat. The entire process, from the touching of the trigger hair to the closing of the door, is incredibly fast, as little as two-thousandths of a second!

Once inside, the prey is digested, mostly by enzymes secreted by special glands but often also aided by bacteria. When digestion is complete, other special cells in the sac wall pump out the water, restoring the concave shape and reestablishing low water pressure (a partial vacuum) inside. The trap is now ready to spring again. The digestive phase of the operation may take as little as fifteen minutes or as much as two hours, depending on the prey's size and digestibility.

If the intricate apparatus of the bladderwort trap were larger and more visible, bladderworts would doubtless have drawn as much attention as the Venus's-flytrap. But so tiny are the traps and so out-of-the-way their homes that this marvelous plant with its intriguing predatory sacs goes largely unnoticed. But for those who seek these carnivores, these animal-eating plants, there awaits one of the ecological and evolutionary wonders of our natural world.

9 | THE ORCHIDS

Paradoxically, some of the rarest plants in temperate North America belong to one of the largest and most widespread groups of plants on earth. A huge family, the Orchidaceae has an estimated 15,000 to 30,000 species, accounting for five to ten percent of all flowering plants in existence. The vast majority of the family is tropical, arboreal, and epiphytic.

Approximately two hundred species occur in North America, most of them terrestrial. Despite their probable tropical ancestry, some species live almost as far north in this continent as there are plants, and many are boreal only.

Of the nearly sixty orchid species known from the Northeast, more than half have been found in peatlands. Sparsely treed, acidic bogs have long had the reputation of being the best places to go to find wetland orchids, but the greatest wealth of species is in wooded and herbaceous fens. At some large peatland complexes, where habitats vary greatly in wetness, nutrient status, acidity, and shade, there may be many species: from the well-studied Orono Bog near the University of Maine some 22 species are recorded from peatland and near-peatland habitats.

Several species could be called cosmopolitan in the Northeast, for they are found from north to south and from the Atlantic to the Great Lakes. Though sometimes abundant in certain peatlands, none can be considered truly common (at least in the sense that maples or dandelions are common). Some species, such as moccasin flower (*Cypripedium acaule*), prefer the shades of softwood-forested peatlands and uplands. Other species, such as the well-known and much-enjoyed grass pink (*Calopogon tuberosus*), rose pogonia (*Pogonia ophioglossoides*), and dragon's mouth (*Arethusa bulbosa*), find the open and semi-open sphagnum/sedge mats more inviting. Occasionally these three species inhabit the same peatland, though grass pink is usually the most frequent. Dragon's mouth is least commonly encountered, especially southward in the region. Among white-flowered orchids, white bog orchis (*Platanthera dilatata*) and white fringed orchis (*P. blephariglottis*) are other successful and more or less common peatland plants.

Other orchids are more limited geographically and/or are restricted to certain peatland types. Southern twayblade (*Listera australis*) approaches its northern limit in central New Jersey, with only a few outposts northward in New York and Vermont; calypso (*Calypso bulbosa*), a rare orchid of northern white cedar swamps and peatlands in northern New England and New York, grows only where the cedar does.

Some species are very rare indeed. The small white lady's slipper (*Cypripedium candidum*) survives at scattered sites, particularly calcareous wetlands in New York and the Midwest, and it was recorded earlier in a couple of places in New Jersey and Pennsylvania. It is currently on the federal list of threatened species. Snowy orchis (*Platanthera nivea*) is restricted to peatlands in the southernmost peatlands of the New Jersey Pine Barrens and is listed by the state as an endangered species. The prairie fringed orchis (*P. leucophaea*) grows in the Northeast at a single fen in northern Maine. Ram's-head lady's slipper (*Cypripedium arietinum*), rare throughout its range, grows in at least one Northeast peatland in southern Maine.

Throughout history, orchids have captured human fancy—

they have been studied, painted, photographed, collected, and cultivated. An almost mystical aura grows around the sound of their name. From Australia's subterranean saprophytes and the tropics' epiphytes to the Northeast's bog and fen dwellers, orchids display remarkable adaptations, especially in their often showy and always fascinating flowers.

In the Orchidaceae sexual organs reveal considerable evolutionary modification. Typically, in each flower the solitary stamen (rarely, species have two or three) is fused to some degree with the pistil. In most orchids the fusion of these male and female parts is complete, the entire structure titled the "column." The pollen formed by the stamen, usually amassed in a large glob (the pollinium), is transferred in its entirety during pollination. Usually intricate, the petals show many exquisite adaptations indicating a long and complex relationship with insects. Commonly, one petal is enlarged or modified into either a lip (labellum) or, as in the lady's slipper, a large pouch. Both formations seem to be designed for insects: the lip acts as a landing platform and the pouch as a non-lethal trap. Both ensure pollination.

The highly evolved orchid flower is the supreme expression of how to attract insects for pollination, for nearly all orchids are utterly dependent on insects for their perpetuation. Unlike many other groups of flowering plants, in orchids the pollen is locked away in wads of pollinia and hidden deep within or protected by the flower itself. Winds cannot bear it away, and even most insects cannot reach it. The pollen can be removed only by sticking to a very specific insect pollinator, thus hitching a ride to another flower. In addition to often being showy, colorful, or strongly scented to attract the appropriate species, the orchid flower typically requires a complex behavior by the pollinator and has developed a floral anatomy to match the body size and shape of the specific insect. Thus, orchids have gone to great lengths to make sure cross-pollination occurs properly. As Dowden (1975) observed, sometimes these lengths almost seem absurd:

The methods which accomplish this [pollination] are often so intricate and fantastic as to seem a joke of nature. Some tropical members of this big family subject their insect pollinators to preposterous indignities—intoxicating them, nearly drowning them in petal-cups of liquid, catapulting them against the pollen.

The orchids in northeastern peatlands do not employ quite such bizarre methods, but certainly they do their share of luring, deceiving, and entrapping to achieve their ends. A couple of species will serve as examples.

The common moccasin flower or pink lady's slipper (*Cypripedium acaule*) has a deep pouch (the slipper), containing the column and many sweet-smelling hairs. A bee—the usual pollinator—enters through a slit in the top of the pouch, perhaps associating the fragrance with nectar, which the hairs do not have. Once it finds itself deceived, it turns to leave, but the only visible exit is not by the entry slit but at the top. Departing, it must crawl past the column. In doing so it rubs against the pollinium, which sticks to the bee. After the bee has been deceived by a second flower and is leaving it, the pollinium strikes the stigma, where it becomes stuck even more firmly by a special stigmatic glue, thus completing the bee's role in the cross-pollination.

The delicate grass pink (*Calopogon tuberosus*) uses another means to the same end. In this species the lip is opposite from the normal orientation—it projects straight up instead of down. The lip is hinged and equipped with many hairlike filaments that to approaching insects must appear to be stamens, indicating the possibility of nectar nearby. However, deception again, for grass pinks have neither stamens nor nectar on the lip. When the visitor (usually a bee) lands on the lip in anticipation of a reward, its weight causes the lip to fold down at the hinge until the bee falls backward into the curved cradle of the column. There the pollinium sticks to the insect's back. The bee struggles out, flies to another flower, and goes through the same series of events but this time deposits the pollinium on the stigma of the column. For cross-

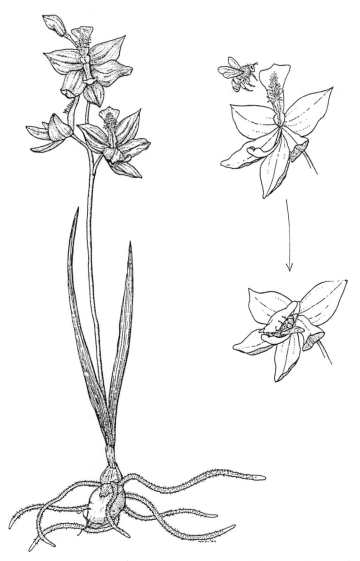

Fig. 45 *Grass pink* (Calopogon tuberosus). One of the more frequently encountered bog orchids, widespread throughout the Northeast, the grass pink comes in various shades of red, from faint pink (almost white) to a deep purple. The "inverted" flower depends on carpenter bees to carry out cross-pollination. The details on the right show the flower with the upper lip or labellum (*top*), the column (*center*), and a bee acquiring pollinium on its back (*bottom*). The intricate steps in this cross-pollination are described in the text.

pollination to be successful the pollinator must be the correct size and weight—heavy enough to cause the lip to fold down and long enough to rest in the column's cradle. Carpenter bees (family Apidae), who have just the right weight and length, are the usual pollinators of grass pink.

At full development, orchid seeds are numerous, light, and tiny—small enough, in fact, to be windborne like pollen. They are among the smallest of all plant seeds, dust size, often numbering thousands per flower.

Orchid seeds are small mainly because they lack endosperm, the food that is normally stored for use by the developing embryo. Instead, the initially poorly organized orchid embryo obtains its food from a fungus that invades each seed. The fungus secures nutrients and transforms them into sugars, which become available to the embryo. Because the seed cannot even germinate without this association, scientists have speculated that orchids are parasitic on the fungus. The association, though apparently less crucial for the orchid, continues in the adult orchid, with the fungus taking up residence in the roots.

It is a grand twist of evolution that the two groups of most highly evolved herbaceous plants, orchids and grasses, have arrived at two very different solutions for pollination and seed dispersal. The orchids depend on insects (and in a few cases birds) to transfer their thick globs of pollen, whereas their seeds are tiny and carried by the wind. Grasses, on the other hand, have minute and light pollen that is windblown but thick, heavy seeds that rely on dispersal from birds, mammals, or other animals that eat them or inadvertently carry them away.

Besides, or maybe because of, such precise requirements for pollination and seedling development, orchids are extremely sensitive about their living conditions. They are also notoriously unpredictable about blossoming even when growing in the most suitable places. For reasons that are not at all clear, a species (such as grass pink or rose pogonia) may almost inundate a peatland one year, and yet the next year not a flower is to be found. It is even more unnerving when after years of absence a species unexpectedly flourishes again in an old haunt.

No plants are more sought, more photographed, more written about, and more enjoyed in the Northeast's peatlands than orchids. Neither soggy underfooting nor swarms of attacking insects deter avid orchid lovers. Many are the natural areas and preserves set aside primarily to protect these most special of our plants.

However, because of their beauty, delicacy, and rarity in the north-temperate lands, orchids also have been the unfortunate targets of collectors. Many peatlands have suffered relentless plundering. The most devastating and long-lasting effects are caused by some commercial collectors who sell wild orchids without thought of conservation. Consequently, it is a real risk to reveal the location of peatlands containing rare species or abundant populations of showy species.

An even greater threat is the outright loss of peatlands in which the orchids live, whether through peat mining, draining, filling, development, or other preemptive operation. Collecting in theory can be controlled to some extent, and the collector seldom destroys the physical setting in which the orchids grow. There is no remedy or hope, however, when their very homes are gone.

10 | SEDGES AND HEATHS

SEDGES

More ubiquitous in peatlands than even sphagnum mosses are the sedges (family Cyperaceae). The Cyperaceae is a huge family of highly evolved monocotyledons, with plants similar in gross appearance to grasses yet significantly different in floral and vegetative anatomy. Sure identification depends on flower and seed structure, but many sedges can be identified as such by their triangular, solid stems (grasses and rushes usually have round and sometimes hollow stems) with the leaves attached directly to the stem instead of forming, as do the grasses, a sheath around it.

The Cyperaceae is the second largest family of flowering plants in the Northeast (after the composites—Asteraceae). No one has enumerated the sedges from the region's peatlands, but at least 150 species from 10 genera are recorded as growing in bogs and fens. All or most of the species of some genera, such as cotton grass (*Eriophorum*) and beak-rush (*Rhyncospora*), occur in peatlands. At least one-third of the species of the mammoth genus *Carex* (the largest genus of

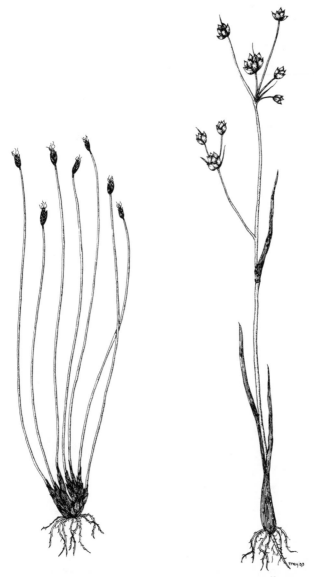

Fig. 46 *Sedges*. Deer's-hair sedge (*Scirpus cespitosus*) (*left*) and twig rush (*Cladium mariscoides*) (*right*) represent the two extremes of peatland preferences: deer's-hair sedge form lawns on Maine's coastal raised bogs; twig rush inhabits rich, calcareous wetlands. Deer's-hair sedge often grows to 3 feet or more, while twig rush reaches less than half that height.

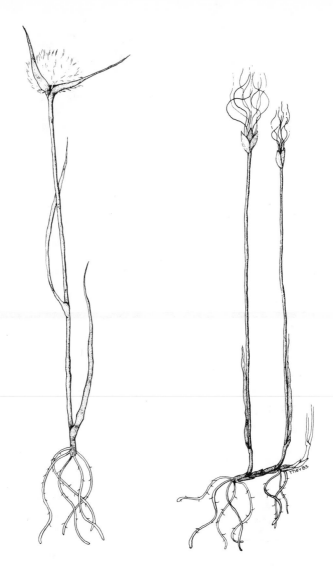

Fig. 47 *Sedges*. Two genera of sedges—cotton grasses (*Eriophorum* spp.) and some bulrushes (*Scirpus* spp.)—have species with cottonlike tufts when fruiting and impart an attractive grace to the places where they grow. The two species shown here, tawny cotton grass (*E. virginicum*) (*left*) and Hudson Bay bulrush (*S. hudsonianus*) (*right*), often grow together. Tawny cotton grass's full tufts are white at the edge and brownish where attached to the stem; the less dense bulrush tufts are of light, white filaments, smaller and more delicate.

flowering plants in the Northeast) grow in peatlands; some are major peat formers in certain fens.

With so many species it is no wonder that in the Northeast there are rare species (in Maine *Carex rariflora*, the scantily flowered sedge, is known from a single site), geographically restricted species (sheathed sedge, *C. vaginata*, for example, extends south only to northern New York, Vermont, New Hampshire, and Maine), and cosmopolitan species (such as *Rhyncospora alba*, white beak-rush, and *Eriophorum virginicum*, tawny cotton grass).

Overall—and especially in the genus *Carex*—sedges present many difficulties for identification. This is unfortunate, since they are excellent indicators of site conditions and often diagnostic of peatland type. Nonetheless, some genera and species are relatively easily learned. For example, the cottony tufts of cotton grasses (*Eriophorum* species) make the genus easy to spot when in fruit. Its species differ in height, number of fruiting stalks, timing of fruiting, and color of tufts. On the other hand, it is the prominent rows of leaves of three-way sedge (*Dulichium arundinaceum*) that make it distinctive. Even among the troublesome *Carex* species a few can be more or less readily identified—the sparse, tasseled terminal cluster of few-flowered sedge (*C. pauciflora*) is quite recognizable even amid other sedges.

Some of the more common species, such as beaked sedge (*C. rostrata*), few-seeded sedge (*C. oligosperma*), water sedge (*C. aquatilis*), hairy-fruited sedge (*C. lasiocarpa*), bog sedge (*C. exilis*), twig rush (*Cladium mariscoides*), and deer's-hair sedge (*Scirpus cespitosus*), however different they appear when contrasted by an expert, remain for most novices difficult at best and essentially impossible to identify when sterile.

Sedges grow in a wide range of wet conditions, but most species are fairly choosy about acidity, dissolved oxygen, nutrients, and shading. Most peatland species live in the minerotrophic waters of fens, though several inhabit ombrotrophic bogs. In fact, by their presence many sedge species tell us what kind of peatland we are in, whether bog, rich fen, or something intermediate.

They also can indicate nutrient or pH gradients within a peatland, often quite strikingly by changes in color or texture from one part of the peatland to another. In the Northeast, for example, abundant hare's-tail cotton grass (*Eriophorum spissum*), few-flowered sedge, or few-seeded sedge denote bog or nutrient-poor waters. Slightly enriched sites may have bog sedge or the mostly submerged bulrush *Scirpus subterminalis*. Peatlands somewhat richer in nutrients might have tawny cotton grass, green-keeled cotton grass (*Eriophorum viridi-carinatum*, which also grows in limy wet meadows), or three-way sedge. In the richest fens of the Northeast one may find twig-rush or Hudson Bay bulrush (*Scirpus hudsonianus*).

Some species, such as white beak-rush and tussock sedge (*Carex stricta*) have wide tolerances, being found from raised bogs to the flowing waters of fens. Others are much restricted; sheathed sedge, for example, in the Northeast apparently grows only in white cedar swamps from Maine to northern New York.

In open fens, the sedges of peatlands reach their fullest expression and diversity. Typically growing in dense tussocks or more loosely as lawns, their shallow and entwining roots and underground stems (rhizomes) form a well-woven mat that, where waters flow, keeps the plants from washing away and provides a base on which other, less firmly rooted plants (such as most mosses) can grow. Where vegetation successionally is encroaching on pond, lake, or stream water, the resilience of a sedge mat gives a framework for subsequent colonization and growth by sphagnum. The dense growth habit of sedges can result in a firm, dense, fibrous peat—obvious enough to anyone who has walked on a pure sedge mat and to European peat cutters working by hand, straining to hack their way down through tough sedge layers to get to more easily cut peats.

Curiously, many sedges seem to reverse our notion that the roots of plants are merely aids supporting and watering the "real" aboveground plant. With so many sedges, the leaves live but a few weeks or months at most, producing carbohydrates for subterranean growth and seed production, then sacrificing themselves as their own nutrients are drawn into the overwintering, perpetuating roots and rhizomes. Even in summer the roots and rhizomes usually contain much more biomass (living material) than do the aboveground leaves. Thus, what is underground is much more the persistent, "real" plant.

Sedges have great visual appeal. In some areas they stand like the high grasses of prairies, waving under the winds, and in the fall almost glowing in russets and golds and tawny browns. The seed heads of the cotton grasses (which appear in the spring for some species, in the late summer or fall for others) often create spectacular displays as the white or tawny tufts mature in unison, giving the peatland the appearance of a ripened cotton field or a meadow of dandelions gone to seed. Reflected in the quiet waters of a shallow bog pool, the little starburst of white beak-rush seeds can appear as miniature constellations above a sea of sphagnum and red-tinted sundews, stunted pitcher plants, and ripened cranberries. Especially pleasing, and much photographed, are the inflorescences of sedges. Some are robust, sturdy, and erect; others are pendants that sway gracefully in the breeze; still others are almost crystalline as early morning dew glistens the pointy seeds that comprise each fruiting head.

Despite the soggy terrain in which they grow and the low-quality fodder their leaves provide, some peatland sedges were used in the Northeast as livestock feed until as recently as twenty to thirty years ago. Pasturing was done (with varying success, one suspects) on all kinds of peatlands, and haying was carried out at firmer, sedge-rich sites. As hand-cutting was replaced by machine harvesting, only periodically dry, clayey peats (such as in some tidal marshes)—those peats that could carry the equipment's weight—were still occasionally hayed. Today the grazing and harvesting of sedges is rare, but the evidence survives: far out in a New Jersey tidal marsh the rotting relic of a long-forgotten horse-drawn hay wagon lies half overgrown, while in York County, Maine, on graying wind-worn posts now deep in peat, rusted barbed wire clings tenaciously to an abandoned purpose.

HEATHS

Just as sedges are the hallmark of fens, heaths are that of bogs. The group is largely boreal in distribution, successful on the nutrient-poor soils of northern latitudes, higher elevations, and peatlands. A number of species grow southward in the Appalachians to the Carolinas or Georgia; three even extend into the milder Atlantic Coastal Plain as far south as Florida and Texas.

The heaths (family Ericaceae—excluding the wintergreens, which some authors incorporate in this family) are represented in the Northeast by some 52 species in 24 genera. Nearly half (at least 24 species in 9 genera) grow in peatlands. All woody shrubs (though some may not at first glance appear so), they range from the prostrate strands of small cranberry (*Vaccinium oxycoccus*) to the near-trees of rhododendron or great laurel (*Rhododendron maximum*), which may reach 30 feet in height while forming virtually impenetrable thickets.

Some fens too have abundant heaths. The most ubiquitous (perhaps the most frequently encountered peatland plant in the Northeast north of New Jersey) is leatherleaf (*Chamaedaphne calyculata*), which thrives in open peatlands of diverse water quality, depth, and flow. At some pond margins it may be so dense and vigorous—upward and pondward—that formerly functioning roots and stems, though still connected, sink deeper and deeper, making massive woody tangles several to many feet below the surface. Highbush blueberry (*Vac-*

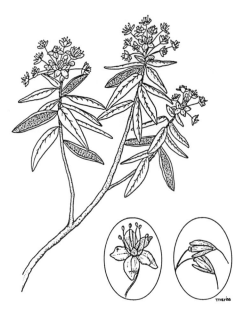

Fig. 48 *Labrador tea (Ledum groenlandicum)*. So named because of its use as a beverage by backwoodsmen (very potent, however), Labrador tea is one of the most common heaths of the north. It grows not only in peatlands but in many upland ecosystems as well, in boreal and subarctic regions. The undersides of the leaves (whose margins curl inward) are covered with white, woolly fuzz, which turns brown as the seasons progress. Prior to fertilization the five white petals of the expanded flowers (*left inset*) form showy terminal clusters; later the petals reflex, the stamens fall off, and the styles remain—all creating little "rockets" (*right inset*).

cinium corymbosum) and black highbush blueberry (*V. atrococcum*), especially in the deciduous forest regions of the Northeast, can so dominate a peatland with their multiple stems, profuse branching, and height (a canopy often 10 to 15 feet high) that passage without the aid of deer trails is an arduous adventure, at the very least!

All the peatland heaths of the Northeast are woody shrubs. Ten bear leaves year-round: Labrador tea (*Ledum groenlandicum*), sheep laurel (*Kalmia angustifolia*), bog laurel (*K. polifolia*), bog rosemary (*Andromeda glaucophylla*), leather-

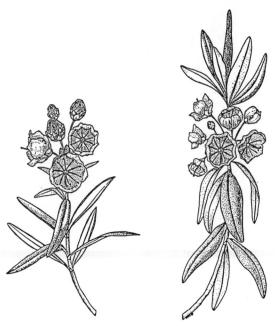

Fig. 49 *Laurels (pale or bog laurel)* (*Kalmia polifolia*) (*left*) and sheep laurel, or lambkill (*Kalmia angustifolia*) (*right*), are two of the more common low shrubs of northern peatlands. Sheep laurel grows in a wide variety of habitats, ranging from wet peatlands to acidic uplands; bog laurel is largely confined to peatlands. The parasol-shaped flowers are similar, but their location on the stem distinguishes the species—terminal inflorescences on bog laurel (which is also usually the smaller of the two) and along the stem in sheep laurel.

leaf, creeping snowberry (*Gaultheria hispidula*), mountain cranberry (*Vaccinium vitis-idaea*), large cranberry (*V. macrocarpon*), and small cranberry. In addition to their apparent resistance to desiccation or damaging winds and ice, the tough, firm, and even scleromorphic leaves of many heaths appear to have other survival traits. In particular, they also happen to be generally repugnant to herbivores and immune to potential infections by some bacteria and viruses. Browsing animals tend to avoid species such as bog rosemary, leatherleaf, and Labrador tea, apparently not so much because of unpalatability but because of a toxin (acetylandromedol—formerly called andromedotoxin—after the scientific name for bog rosemary) produced by the plants. The leaves resist infection by microbes through production of tannins and other organic substances. Moreover, the plants not only create these chemical barriers to invasion but also seem to be able to neutralize pathogenic enzymes.

To bolster the conservation of nutrients achieved through evergreenness, heaths maintain symbiotic dependencies with root fungi, similar to that of orchids. This mutually beneficial mycorrhizal association, as it is called, occurs in most heath species. Perhaps there is a connection between the mycorrhizal habit and the prolific berry crops produced by many of the heaths, particularly the blueberries, cranberries, and huckleberries (those previously mentioned plus bog bilberry, *Vaccinium uliginosum*; dwarf huckleberry, *Gaylussacia dumosa*; and two medium-size blueberries of the Coastal Plain, *V. caesariense* and *V. marianum*).

Two of the heaths produce well-known berries under extensive cultivation (see Chapter 15). The firm, red berries of large cranberry become cranberry sauces, preserves, and a wide variety of beverages. Berries of low sweet blueberry (*V. angustifolium*) find their way into many pies, tarts, muffins, or fruit salads. Commercial cranberries primarily are raised on peatlands modified with drainage controls and mineral (usually sand) additives. Southeastern Massachusetts produces nearly all of the Northeast's commercial cranberries. Blueberries are grown for harvest principally on uplands, though peatlands provide many bucketloads for home use, and in past times a significant number of peatlands were periodically burned to enhance blueberry production. Southeastern Maine and southern New Jersey have the largest commercial operations, but small blueberry fields can be found scattered here and there throughout the Northeast.

Biologically, heaths clearly are an important component of most peatlands, in terms of sheer numbers of plants, their

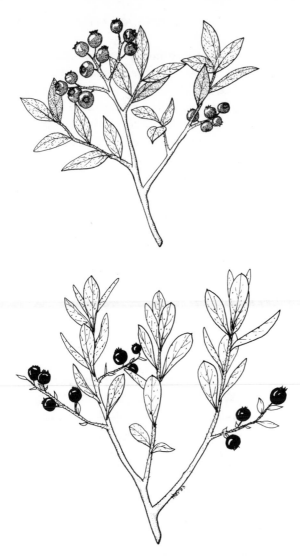

Fig. 50 *Blueberries and huckleberries*. Velvetleaf blueberry (*Vaccinium myrtilloides*) (*top*) and dwarf huckleberry (*Gaylussacia dumosa*) (*bottom*) are only two representatives of many "berry" heaths in peatlands. The berries of velvetleaf blueberry are a soft azure, whereas those of huckleberry are deep purple to almost black. Huckleberry leaves turn a beautiful orange-red in the fall, setting "afire" many a raised bog.

contribution to overall productivity and peat accumulation, and the genetic value of species diversity. But additionally, they provide an unmeasurable aesthetic dimension. Sweet is the fragrance of many blueberries' ethereal, pendant flowers. Pleasant and sometimes showy are the blossoms of laurels and rhododendrons. Yet perhaps most appealing are the changing colors of the seasons brushed across open peatlands by waist-high shrub forests of heaths. Surely no place can be more beautiful than a bog in full display on a clear fall day: blueberries in shades of red, the scarlet and crimson huckleberries—so intense as to seem to be giving off light of their own—the soft purples of leatherleaf, rhodora, and bog rosemary. Even in winter, the etching of a twisted stem, persistent leaf, or solitary berry is a gentle image of lives beneath the snow. And in early spring, the delicate flowers fulfill that earlier promise.

11 | INSECTS AND OTHER INVERTEBRATES

Insects may well be the most abundant and various of all peat-land animals, at least those that we can see without the aid of a microscope. They occur in the waters, among sphagnum, and on and about the herbaceous plants, the shrubs, and the trees. Collectively, insects no doubt provide the bulk of herbivory in peatlands, and invertebrates as a whole are important decomposers in the aerobic zone.

The aquatic insects of peatlands are typical of slow-moving or stagnant waters in other wetlands, streams, ponds, and lakes. Curiously, even in bog waters whose extreme acidity and low nutrient content make them the most distinctly unique of peatland waters, there are few if any endemic insects. To be sure, species' abundances and diversities are much less than in fens and marshes, but what seems remarkable is that insects such as the common water boatmen (family Corixidae) and whirligig beetles (family Gyrinidae) can survive in bog waters with pHs as low as 3.5!

Despite their motility, a number of flying insects spend all or part of their lives in peatlands. Many are northern tundra or boreal species, finding suitable habitats southward only in peatlands or alpine heights, just as do several biogeographi-

Fig. 51 *Jutta arctic* (*Oeneis jutta*). This dusky brown butterfly is an inhabitant of far northern peatlands around the world and provides moments of excitement for the collector. "When you have chased one through a bog, sinking to your knees at every step in saturated sphagnum moss, hurdling small tamaracks and black spruces, and dodging around larger ones, tripping over clumps of heaths, and boring through a cloud of bloodthirsty blackflies, you have earned your specimen—if you can catch it" (Klots 1951). The butterfly has an approximate wingspan of 1 inch.

cally similar herbs and low shrubs. Several species of rather inconspicuous butterflies are examples: the jutta arctic (*Oeneis jutta*), the bog and lesser purple fritillaries (*Boloria eunomia* and *B. titania*), the bog elfin (*Incisalia lanoraieensis*), and the bog copper (*Lycaena epixanthe*). As might be expected, the habits of these butterflies are tied closely to the food sources of the larvae (caterpillars) or adults. Bog copper caterpillars dine exclusively on the leaves of the large cranberry (*Vaccinium macrocarpon*); those of the bog elfin eat the tender new needles of black spruce (*Picea mariana*). The elfin also drinks nectar from the flowers of rhodora (*Rhododendron canadense*), flying back to the spires of black spruce to rest. Both species lay their eggs on the food plant—bog coppers on cranberry leaves, bog elfin on spruce needles.

The most intricate relationships between peatland insects and plants involve but a single species of each. One of the most fascinating exists between a small moth, known by its Latin name, *Exyria rolandiana*, and the pitcher plant (*Sarracenia purpurea*).

The adult moth lays its eggs in a pitcher plant leaf. On hatching, the caterpillars crawl to other leaves on the same plant, distributing themselves so that each pitcher has only one caterpillar. In species of pitcher plants with isolated, single leaves (none of which grows in the Northeast) another *Exyria* species distributes its eggs differently—one per leaf. Thus, the winged parent solves a difficult or impossible distribution problem for the caterpillars. For *Exyria*, then, the growth habit of the food plant has determined the evolution of the egg-laying habits.

The spring-hatched caterpillars of *E. rolandiana* feed on the developing interior lining of the pitcher. While there they may spin a web across the mouth of the pitcher to protect themselves from predators. They may even "girdle" the tissue below the hood, causing it to fall over the opening and make the pitcher into a safe, dry place. As fall approaches, the encapsulated leaf becomes a ready-made cocoon for the overwintering pupae.

The *Exyria* caterpillars have other adaptations that show how closely they have evolved with pitcher plants. Some have colorations very much like the interior linings of their host plants, the deep reds often resembling the veins of the plants. The caterpillars of two species have several bumps (lappets) on their sides (which are lacking in *E. rolandiana*), believed to enable the caterpillars to enter the very narrow pitcher of the small pitcher plant (*S. minor*) without getting stuck. In the wider leaves of *S. purpurea*, *E. rolandiana* caterpillars face no such danger and thus have not developed lappets.

Fig. 52 *Pollinator predator.* A crab spider (family Thomisidae) resembles a flower of the white bog orchis, *Platanthera dilatata* (see the lowest "blossom"), as it waits, forelegs outspread, to capture an unwary insect. As the author moved around the plant to take this photograph, the spider moved slowly around the stem, using its best camouflage at all times. *(Photo by Charles W. Johnson.)*

Another use of plants by insects, one readily encountered in peatlands, involves predators cashing in on the attracting powers of plant flowers, fruits, and even leaves (such as those of pitcher plant). Several predatory insects hide among plants, stealthily waiting a chance to grab prey as it flies, crawls, hops, or walks by. The best known are the "pollinator predators" that lie in wait in or on orchids. Adept at such fatal surprise tactics are species of ambush bugs (family Phymatidae), assassin bugs (family Reduviidae), and crab spiders (family Thomisidae). They may hide inside the flowers (one group of spiders actually lives within the pouches of lady's slippers) or stay outside and mimic the flowers' shape or color. Like chameleons, some species can even change color to match various surroundings. The predator-orchid relationship is specific, for the predator relies on pure deception to get its meals and must match its background exactly or hide extremely well.

A visit to a peatland can reveal many other insects, spending at least some part of their lives there, from the whirring dragonflies and damselflies to the intently scurrying ground beetles to the noisy but reclusive grasshoppers and crickets. Among several interesting surprises, I once encountered a huge nest of white-faced hornets (*Vespula maculata*) constructed within the multiple stems of a leatherleaf bush, perfectly hidden by the leaves. Black and even red ants (family Formicidae) colonize firmer hummocks or decaying wood. In open peatlands they favor hummocks of tough, rapidly growing bog haircap moss (*Polytrichum strictum*); in alder peatlands, during episodes of high water, ants will make mounds of peat progressively higher to escape flooding and protect the colony.

The invertebrate fauna of peatlands extends beyond insects, of course. A large proportion are detritivores that feed on dead plant and animal material in various states of decomposition. They live primarily in the thin aerobic zone where detritus accumulates but is not perpetually saturated (or nearly so), since few can survive the rigors of waterlogging and lack of oxygen of the lower peats. Oxygen is crucial for the invertebrate decomposers, for it is the combining of oxygen with broken-down carbohydrates that provides the major part of their bodily energy. Earthworms (family Lumbricidae) and millipedes (order Diplopoda), for example, are rare, if they are present at all, in saturated peats. (Most of us have seen earthworms covering lawns and sidewalks after soaking rains, fleeing to the surface as they seek oxygen.)

Many soft-bodied invertebrates appear intolerant of high acidity and thus cannot live in bogs, but in fens they may be able to function quite well. This is the case with certain kinds of earthworms, which are absent from bogs but present in some fens where water conditions are also suitable. Freshwater clams (class Pelecypoda) too may be found in fens but not in bogs. The bog waters preclude development of the shells, which, being made of calcium, are highly soluble in such acid conditions.

Several predatory groups, such as spiders (order Aranae) and centipedes (order Chilopoda), are well represented. Various mites (order Acarina), parasitic on insects or other animals, are often present in abundance. Invertebrates that feed on live plant material are more evident, too. Slugs and snails (both class Gastropoda), for example, are quite common in bogs, as can be seen from the many slime trails they make in their nocturnal meanderings, often highlighted by an early morning dew. (Slugs likely are a staple in the diet of other peatland animals in bogs, such as the small redbelly snake *Storeria occipitomaculata*.)

No doubt future research on invertebrates, including those too small to see with the naked eye, will expand our understanding of the roles they all play in making peatlands work. The surge of research during the last few years on peat decomposition is leading to studies on the decomposing agents themselves. Exactly how conditions in different kinds of peatlands encourage or discourage certain invertebrates is a subject of a number of current investigations. Their findings should open new insights about nutrient recycling, peat accumulation, food chains, and the overall ecological dynamics of peatlands. There is indeed much peatland wildlife yet to see if we pause to look closely enough.

12 | FISH, AMPHIBIANS, AND REPTILES

FISH

Peatlands, especially bogs, do not have a reputation as good fishing spots, and for the most part that reputation is deserved: the waters are too acidic, too stagnant, too devoid of oxygen, too meager in food to support great numbers and kinds of fish. Only where the Northeast's peatlands are adjacent to ponds, lakes, rivers, or bays and estuaries of the Atlantic do they occasionally play significant, if sometimes indirect, roles in supporting fish populations.

Atlantic salmon (*Salmo salar*), for example, spawn in Maine's Pleasant River, whose headwaters are in a huge peatland—the 4,000-acre Great Heath. Since the salmon require unpolluted water in which to live and reproduce, the undisturbed peatland probably is important in maintaining proper water quality (a matter of great concern, considering the impact of peat mining). But beyond its role as protector, the peatland seems to provide nothing unique for the salmon's needs.

Acidity is a major reason fish are unable to survive in the

ponds, pools, and streams of bogs and many fens, for low
pHs are toxic to fish in several ways. In water with a pH below 6.0 the eggs of most species become severely deformed;
below a pH of 4.0, eggs fail to develop. The acidity also
impairs calcium metabolism of the adults and the oxygen-
absorbing capabilities of their gills.

Because of the higher alkalinity, certain fens (including
laggs of Maine's raised bogs where there are flowing streams
or beaver ponds) do have fish populations. Fens adjacent to
lakes, larger streams, and rivers may even support substantial
game species. Some of Maine's large northern fens, for example, have such cold-water species as landlocked salmon
(*Salmo salar*), brown trout (*Salmo trutta*), lake trout (*Salvelinus namaycush*), and brook trout (*Salvelinus fontinalis*)
and some warm-water species such as largemouth bass (*Mi-
cropterus salmoides*) and chain pickerel (*Esox niger*). The
fish are dependent on clean water; decreased oxygen content,
increased acidity, and influxes of suspended organic particles
from disturbances in peatlands could pose serious problems
to many fisheries.

AMPHIBIANS

Very little research has dealt with the amphibians of peatlands. Although we know that some frogs, toads, and salamanders do live in peatlands (see Appendix 2), we do not
know much about their habits there or their importance in
overall peatland ecology. They may well be more significant
than we have assumed.

A recent preliminary survey of the wildlife of Maine peatlands revealed eight amphibian species, widespread throughout except in forested peatlands. Included were American
toad (*Bufo americanus*), five true frogs (green frog, *Rana
clamitans*; mink frog, *Rana septentrionalis*; wood frog, *Rana
sylvatica*; northern leopard frog, *Rana pipiens*; pickerel frog,
Rana palustris), and two tree frogs (spring peeper, *Hyla cru-*

cifer and gray tree frog, *Hyla versicolor*). All of these species occur in nonpeatland habitats as well, and all but the mink frog (restricted to the northern third of the Northeast) live throughout most of the region. Known from or expected at peatlands south or west of northern New England are a few additional species, notably the northern cricket frog (*Acris crepitans*), Pine Barrens tree frog (*Hyla andersoni*), carpenter frog (*Rana virgatipes*), mud salamander (*Pseudotriton montanus*), and red salamander (*P. ruber*). Some of these might be considered rare, and one, the Pine Barrens tree frog, is listed as regionally endangered.

Acidic and nutrient-poor peatlands, especially bogs, are difficult environments for amphibians. Foraging for and preying on invertebrates can be moderately rewarding, but as in the case of fish, acidity presents problems. Most amphibians lay their eggs in water. The eggs, with only gelatinous covering for protection, are very sensitive to acidity, suffering serious defects even in mildly acidic environments. For most species a pH of 4.0 or lower is lethal. The wood frog, whose range extends to the arctic circle and beyond and is thus one of the few truly boreal amphibians, is one species that seems more or less immune to the acidity of bog waters. Another northern species, the mink frog (named for the musky, mink-like odor it emits on being handled) is also probably immune, since its range includes much of Canada, the northern United States, and the higher montane elevations in the East—environments abundant in acidic waters.

No amphibian species lives only in peatlands, but a few can be regarded as predominantly peatland species. One of the most faithful to peatlands and a rather uncommon species is the handsome four-toed salamander (*Hemidactylium scutatum*). Its name derives from its most distinguishing feature, four toes on the hind feet instead of five, as all other salamanders in the region have. The four-toed salamander is active within the moist sphagnum hummocks and lays its eggs in pools or other nearby water bodies. Because it spends

Fig. 53 *Four-toed salamander* (*Hemidactylium scutatum*). The four-toed salamander, usually considered rare, in some areas may be more common than we think. It is quite secretive and spends much of its life hidden below the ground surface. In peatlands it tends to remain within moss hummocks, not too distant from pools or ponds. Eggs are laid in moss but next to water. The female guards the eggs until they hatch; the larvae eventually go to the water, where they stay for a couple of years while maturing. As adults, they move back to land. The species is easy to identify: it has four toes on the hind feet, a light yellow or red back, a white belly with black spots, and a constriction at the base of its tail. Adults measure about 3 inches long (including the tail).

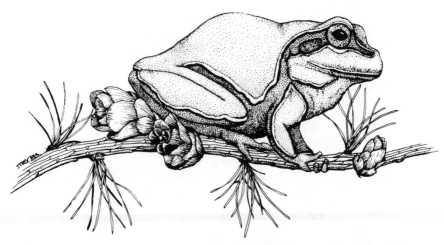

Fig. 54 *Pine Barrens tree frog (Hyla andersoni)*. This most handsome little tree frog, no more than 1.5 inches long, has a pea-green base color and a white-edged, purplish band down its sides. It is restricted to wetlands (including peatlands) from central New Jersey south along the Atlantic Coastal Plain. Already rare throughout its range, it is further threatened by destruction of its wetland habitats.

much of its life beneath the surface, among the mosses and litter, this secretive peatland dweller remains seldom seen.

One of the most attractive of the Northeast's amphibians and one of the rarest in North America is the Pine Barrens tree frog. Living only in small portions of the eastern Coastal Plain (predominantly in the area for which it is named, the New Jersey Pine Barrens), this small tree-climber prefers sedge-dominated peatlands and wetlands around natural seeps or springs with a scattering of trees and shrubs where it can take refuge. Throughout its range (primarily New Jersey, with scattered populations from Long Island to Florida) this tree frog is classified as either endangered or threatened because it is restricted to a limited kind of landscape and has to contend with development continuing to encroach on its precious homeland.

More common but also restricted in the Northeast to the Pine Barrens is another Coastal Plain species, the carpenter frog. A "true frog" (in the same family as the bullfrog *Rana catesbeiana*, green frog, and others mentioned earlier), it is nicknamed the "sphagnum frog" because of its preference for peatlands, where its brown coloration blends well with the waters made dark by humic acids. During the breeding season (March to May) the male carpenter frog's mating call sounds like a hammer hitting nails; hence its common name. But because it is nocturnal, secretive, and aquatic, little is known of the biology of this amphibian.

Fens, diverse in type and with waters of moderate pH, contain a delightful variety of amphibians for the inquisitive naturalist. The aquatic eastern (or red-spotted) newts (*Notophthalmus viridescens*) swim and lay their eggs in fen pools and shallow ponds, whereas their land-bound form, the red eft, may be encountered struggling up and down sedge tussocks, over logs, and through mossy groves. The shallows around fen pools or rivulets may support numerous green frogs, with northern leopard frogs busy in sedge lands searching for insects. As the snows disappear from northern lands, the fens resound with the welcome trill of the spring peepers, awakened from hibernation, vocalizing their territorial claims and calling to prospective mates.

The eastern newt, green frog, and northern leopard frog thrive abundantly in peatlands and many other habitats. But what do we know of less common, more secretive, amphibians—especially the salamanders—in peatland habitats? Very little. Perhaps, as in some pond and upland environments, they fill major roles in food webs as carnivores. Perhaps there are species or varieties yet to be discovered. Whoever lives in the peatlands, whatever they do, there is much yet for us to discover, to understand, to respect, and to wisely protect.

REPTILES

Like amphibians, the reptiles found in peatlands of the Northeast are few in species, low in numbers in the more acidic sites, occur in other habitats, and include a few taxa of considerable interest. Sparse anywhere in the Northeast, lizards totally shun peatlands. Some six species of turtles and eight species of snakes regularly employ one or more types of peatland as their habitat (see Appendix 2).

With few exceptions, the reptiles found in peatlands of the region become less numerous with elevation and "northeastness"—eastern and northern Maine is much poorer in species, for example, than New Jersey. Some of the best-known turtles and snakes, familiar to most schoolchildren as well as to naturalists, are often encountered in peatlands. Perhaps the most ubiquitous is the carnivorous common garter snake (*Thamnophis sirtalis*), a sometime resident of bogs and fens where water tables are slightly below the surface. Painted turtles (*Chrysemys picta*), the most generalist of the Northeast's turtles, favors eutrophic and productive waters where herbaceous plants provide food plus sufficient sunlight for basking. Even the largest freshwater turtle of the area, the snapping turtle (*Chelydra serpentina*), can be found where productive waters a couple of feet or more in depth occur in or border peatlands.

Two other snakes, the redbelly snake (*Storeria occipitomaculata*) and the northern water snake (*Nerodia sipedon*), live in and along many peatlands. Redbelly snakes are most likely to be found under logs, fallen bark, or other cover where invertebrates (especially slugs and earthworms, if available, and insects) provide abundant food. Look for this small (eight to ten inches long), inoffensive reptile in habitats ranging from forests and woodpiles to a variety of bogs and fens, throughout most of the Northeast except for the extreme northern tip of Maine. The redbelly is small, nonaggressive, nonbiting, and beautifully colored with its red underside—

traits that make it ideal for close inspection by children and adults somewhat timid about snakes. Like its cousin the northern water snake, the redbelly does not lay eggs but gives birth to its young.

Unlike the small and placid redbelly or the gentle garter snake, the northern water snake has a generally deserved reputation for being big (2 to 3½ feet long) and aggressive. Typically black or darkly hued, this carnivore sometimes (especially when young) displays banding patterns in attractive shades of browns and tans. Most at home in water, at least when startled or threatened, the water snake frequents peatlands bordering ponds or lakes or peatlands with many pools and channels—those that are apt to harbor many frogs and small mammals. Water snakes spend a lot of time basking, a time when they are most often discovered. When handled, they fight vigorously and may bite with some authority (they are not poisonous, however)—hence the reputation of meanness. But when startled or approached in the wild, in my experience they always flee, sometimes with amazing speed.

The only poisonous snakes one might encounter in the Northeast's peatlands are the massasauga (*Sistrurus catenatus*), or "swamp rattler," whose range extends eastward only into western New York and Pennsylvania, and the copperhead (*Agkistrodon contortrix*), which occurs across central Pennsylvania, southern New York, Connecticut, and southwestern Massachusetts. The massasauga is a very rare medium-size (usually around 2 feet long) rattlesnake of wet meadows, wetlands, peatlands, and even uplands in the Northeast, with a shy and unobtrusive demeanor—characteristics that make it virtually unseen in our region. Copperheads, lazy snakes of woodlands and ledges, sometimes visit or inhabit wetlands, presumably including an occasional peatland. The fabled cottonmouth (water moccasin—*Agkistrodon piscivorus*) is not known in the Northeast; extreme southeastern Virginia is its northern limit.

From extreme southern Maine and New Hampshire south

Fig. 55 *Bog turtle* (*Clemmys muhlenbergii*). This small (3 to 3 1/2 inches long) rare turtle has a discontinuous distribution in the Northeast. Known from scattered colonies in New York, Pennsylvania, and New Jersey, as well as a few locations in Connecticut, Massachusetts, and Rhode Island, it is dwindling over all its range. It is now listed as endangered by most states where it occurs. This semiterrestrial reptile prefers open, sedgy, shrubby wetlands—fens, wet meadows, and shallow marshes. Unlike its close relative, the spotted turtle (*Clemmys guttata*), it can eat without submerging its head. Its bright orange ear patches are distinctive. It is most apt to be seen above ground and out of water in springtime, during its mating season.

through the rest of the Northeast, the spotted turtle (*Clemmys guttata*) inhabits many types of ponds and wetlands, including peatlands with a great deal of open water interspersed with tussocks of sedges or other emergent plants. It is a small (about a four-inch-long shell) but striking species, with a pure black carapace adorned with large yellow dots. It uses the waterways for hunting and swimming and the vegetated areas for basking in the sun and laying eggs. It hibernates comfortably and safely deep in the peaty muds.

The spotted turtle's close relative, the bog turtle (*Clemmys muhlenbergii*), is somewhat smaller (only about 3.5 inches long), more terrestrial in its habits, and considerably rarer. It is, in fact, one of the rarest nonoceanic turtles on the continent and is listed as an endangered species in most states where it occurs. Its range is generally to the west of the spotted turtle's, and it is severely confined to treeless fens, wet meadows, and the occasional very poor fen or bog. By virtue of its strong preference for graminoid, often calcareous, wetlands rather than bogs, a more appropriate name for this rarely met turtle might be "fen turtle."

Housing developments, roads, conversion of wetlands to agricultural lands, and other human preemptions have left fewer and fewer habitats for bog turtles and have increasingly isolated those that remain. This trend means not only that the turtle sites are being eliminated outright but also that possible emigration routes to existing or new wetlands are being cut off. As a result, the few turtles that remain must live out their lives in the now-isolated "islands" where they were born. Even if these places are protected, it is possible that natural changes within the wetlands (such as the growth of shrubs and trees at the expense of preferred emergent vegetation) may eventually make them unsuitable for habitation by the turtles. On top of all this, the bog turtle's rareness and small size have made it an attractive target for pet dealers, but it is hoped that trade in these special animals has ceased. Along with timber rattlesnakes (*Crotalus horridus*) and sea turtles, the bog turtle has to rank as one of the most critically threatened reptiles of the region. It requires more than passive concern; it needs active protection if it is to survive at all in the Northeast.

13 | BIRDS

A surprising number of birds come to the peatlands of the Northeast, either to nest, to eat, for cover, or to rest during migration (see Appendix 3). Many of the species are familiar, a few are uncommon, and a select few are truly rare. Birds that live in or visit peatlands are not attracted so much to the peat substrates (to which they appear oblivious, except for a few pond waters) or the specific vegetation as to the physiognomic appearance of the place. Birds of the open field come to open bogs; birds of the marsh come to fens; birds of the forest come to forested peatlands. Since some peatlands, especially those most nutrient-poor and acidic, have plants and communities prominent in boreal North America, these bogs also attract a number of northern species south of their more typical range.

Thus, in peatlands thick with conifers—notably black spruce and tamarack, for example—there is usually an abundance of warblers and flycatchers (especially in migration) typical of coniferous forests in general. Look for species such as yellow-rumped warblers (*Dendroica coronata*), Tennessee warblers (*Vermivora peregrina*), brown creepers (*Certhia americana*), olive-side flycatchers (*Contopus borealis*), and yellow-bellied flycatchers (*Empidonax flavirentris*).

Similarly, in peatlands with ponds or pools a few water-fowl, particularly puddle types such as the omnivorous mallard (*Anas platyrhynchos*) or pied-billed grebe (*Podilymbus podiceps*), may be found. In herbaceous fens rich in sedges, bulrushes, or cattails expect to find common snipe (*Gallinago gallinago*) or sora (*Porzana carolina*) walking furtively in the tall vegetation in search of food. Listen for marsh wrens (*Cistothorus palustris*) and red-winged blackbirds (*Agelaius phoeniceus*) noisily declare territory, while kingfishers (*Ceryle alcyon*) give their rattling call if streams or lakes abut the vegetation.

In open bogs dominated by ericaceous shrubs and sphagnum mosses the avifauna differs too. Watch for flitting white-throated sparrows (*Zonotrichia albicollis*) and Savannah sparrows (*Passerculus sandwichensis*), birds also typical of drier shrublands and even fields. Raptors such as the red-shouldered hawk (*Buteo lineatus*) and barred owl (*Strix varia*) may seek voles or lemmings and may even nest in bordering trees.

In the large northern peatlands of the New York Adirondacks, Maine, and adjacent Canada (and to a lesser extent in New Hampshire and Vermont) can be seen several boreal species at or near their southern limits in eastern North America, birds much sought after by bird-watchers from more southern states, particularly black-backed woodpecker (*Picoides arcticus*), boreal chickadee (*Parus hudsonicus*), gray jay (*Perisoreus canadensis*), and spruce grouse (*Dendragapus canadensis*). Each of these species favors spruce-fir woodlands, especially where open land is adjacent. Consequently, peatlands provide a most suitable setting for these species that may be the daily companions of loggers but are rarely seen by most of us in the Northeast.

Even for these resident species, peatlands are only part of their total habitat. The spruce grouse, a darker boreal counterpart of the more familiar ruffed grouse (*Bonasa umbellus*, sometimes called partridge) of the northern hardwoods, spends much of the summer in uplands, pecking away

Fig. 56 *Black-backed woodpecker* (*Picoides arcticus*). This large (about 1 foot from head to tail) black-on-white woodpecker inhabits spruce-fir forests in and near many of the large peatlands in northern New England and New York. Its primary food is wood-boring grubs of dead conifers. The male has a yellow crown patch.

Fig. 57 *Spruce grouse (Dendragapus canadensis)*. A common inhabitant of North America's boreal coniferous forests, the spruce grouse is an uncommon and intriguing inhabitant of spruce and fir woodlands in and near some large bogs of northern Maine, New Hampshire, Vermont, and the New York Adirondacks. In winter it stays mostly within the protection of evergreens and eats conifer buds and needles. Well equipped for life in snow country, its legs and feet are completely feathered to the toe. It is also rather tame, or at least unwary, allowing people to approach to as close as arm's length.

at a wide variety of berries, shoots, nuts, and leaves. In the fall it moves into the coniferous lowlands, peatlands included, and begins dining more on the needles and buds of softwoods, a diet that carries the birds through the long, snowy, and often frigid winter. Spring courtship rituals and the nesting take place in the coniferous forests, frequently at or near peatlands. When the young are sufficiently large, they leave the nesting areas, continuing the seasonal cycle as they move to nearby uplands for summer foraging.

The large, open, shrub-dominated bogs of the Northeast do attract other, less common birds—predators such as the day-flying short-eared owl (*Asio flammeus*) and northern harrier (*Circus cyancus*) or songbirds such as the Lincoln's

Fig. 58 *Palm warbler* (*Dendroica palmarum*). The palm warbler builds its nest in relatively dry areas in the bog, even within or under hummocks of sphagnum moss. The nest itself consists mostly of grasses of various kinds. It breeds in the extreme northern part of the United States and in Canada. Look for a red-capped, tail-wagging warbler that spends a great deal of time on or near the ground.

sparrow (*Melospiza lincolnii*) and the tail-wagging palm warbler (*Dendroica palmarum*). Of all the birds that make use of peatlands, the last two are perhaps the most bound to them. Their summer range extends from northern Canada to northern New England and New York, but nowhere are they abundant. Both nest in low shrubs or on hummocks in the peatlands, apparently their only nesting habitats in the Northeast. Neither is totally restricted in its habits to peatlands, of course; the palm warbler, for example, moves to open fields in autumn, and both migrate southward for the winter.

The unexpected do show up in peatlands. I have seen nesting bobolinks (*Dolichonyx oryzivorus*), which normally prefer open fields, in large bogs in the New York Adirondacks and in Maine. Snowy egrets (*Egretta thula*) have been sighted in several coastal peatlands. Even bank swallows (*Riparia riparia*) are bog birds in Maine, where at one seaside peatland, in exposed, wave-cut, vertical banks, they dig their nesting burrows into peats just as they do in silts on the faces of river cuts or the walls of gravel pits.

There remains much to learn about the birds of northeastern peatlands. We understand little of their role in food webs and ecosystem function here—for example, do migrating warblers eat significant quantities of defoliating insects in wooded peatlands, or do tree swallows (*Tachycineta bicolor*) consume vast numbers of potential pollinators as they swoop over open fens and bogs? We wonder if any particular peatlands provide crucial breeding or feeding habitat for rare or threatened species. Perhaps the day will even come when our few species extirpated from some of the Northeast's peatlands may be successfully encouraged to return. What a joy it would be to see again bald eagle (*Haliaeetus leucocephalus*) nest high in trees overlooking our lakeside bogs or tidal fens.

14 | MAMMALS

Before Europeans settled in North America and began wielding their influences on the lands and waters, several large mammals inhabited great areas of North America. In all or part of the Northeast roamed the herbivorous white-tailed deer (*Odocoileus virginianus*), moose (*Alces alces*), elk (*Cervus elaphus*), and caribou (*Rangifer tarandus*), along with the predators grey wolf (*Canis lupus*) and eastern cougar (*Felis concolor cougar*). All likely visited various peatlands; some, notably wolf and caribou, sought peatlands out at times as sources of food.

The encroachments of Western civilization slowly and steadily pushed most of these species to ever more remote parts of the continent. In our region all but the white-tailed deer and awesome moose have vanished, and not so long ago even those were nearly eliminated. Maine, New Hampshire, and Vermont have sustaining moose populations in their more boreal regions, although only Maine has substantial numbers (estimated at some 20,000, in contrast to Vermont's 50 to 100). New York has more than adequate habitat for moose in the wilds of the 3-million-acre Adirondack Preserve, but this grand species has only recently begun to recolonize

Fig. 59 *Moose (Alces alces).* A bull in velvet wades in a boreal forest pond. This awesome animal is a frequent summer visitor of bog ponds and large northern peatlands in the Northeast. *(Photo courtesy of Vermont Fish and Wildlife Department.)*

this former territory. The huge herbivore (weighing from 600 to 1,200 pounds), equally at home in forests, shrublands, and wetlands, browses on peatland shrubs and small trees and wades into ponds and deadwaters to uproot water lilies and all manner of aquatic vegetation. It is the perfect muskeg traveler: long legs allow free and easy wading, wide-splayed feet support the great weight in mucky or unsteady ground, and the long, strong snout permits underwater rooting about for food. In Maine's large bogs, well-worn moose trails mark paths across the open expanses, providing shortcuts and anastomosing networks in broad, marginal laggs, likely used for both transportation and local browsing.

The moose's smaller and more abundant cousin, the white-tailed deer, similarly passes through peatlands on its way to more favored feeding grounds. Numerous interconnected, well-worn trails in shrub-rich peatlands testify to the seasonal abundance of browse. Some forested peatlands, especially in central and southern regions of the Northeast, play larger roles in the lives of deer, as they form part of yarding areas where the deer congregate to spend the difficult times of winter. Although midwinter food is meager in spruce or tamarack peatlands, these evergreens provide shelter from the winds and deep snows. Preferred are fens of northern white cedar (*Thuja occidentalis*) with their greater abundance of more acceptable food.

Stealthy predators of the cat (Felidae) and weasel (Mustelidae) families are, like the moose, occasional visitors to the Northeast's peatlands. They shun the expansive unforested peatlands, where high visibility makes hunting unrewarding. The lynx (*Felis lynx*) and bobcat (*Felis rufus*) and the fisher (*Martes pennanti*) and weasels (*Mustela* species) use peatlands the most, largely restricting their travels to rich, wooded sites: thick shrub fens, laggs, and cedar swamps, where their prey, the small mammals of the understory, are most prevalent.

The lynx's whereabouts are closely tied to those of snowshoe hare (*Lepus americanus*), its favorite and nearly exclusive food. Unfortunately, the lynx is now extremely scarce over much of its original range. In Vermont and New Hampshire it is listed as endangered and may in fact be extinct. Maine is its last stronghold in the Northeast, but even there this elusive wildcat may be in trouble. Thus, peatlands, especially the large ones, may play an important part in the lynx's continued residency, now or sometime in the future.

The fisher, on the other hand, after a century of comparable peril in the Northeast, is now doing quite well. Handsome in its dark fur coat and long bushy tail, this large weasel (eight to fourteen pounds) became increasingly scarce from 1850 to 1950 as a result of persistent heavy trapping and early indiscriminate logging. During this period it became restricted to the more inaccessible sections of country, among them mountains and peatlands.

Thanks to reforestation, controls on trapping, and the efforts of humans to restore it in its original areas and to establish new ones, the fisher has returned to the Northeast. Some of its success is also due to its own nature. It can live in a variety of forests, softwoods and hardwoods alike, and can eat an astonishing array of foods, including small mammals, porcupines (*Erethizon dorsatum*, also a sometime peatland animal), carrion, eggs, birds, and some vegetation. It is now solidly, if not abundantly, established throughout much of the Northeast, including larger and forested peatlands.

The beaver (*Castor canadensis*), another forest-based mammal, has experienced changes in fortune similar to those of the fisher. The beaver's existence in the Northeast has likewise been tied to human demand for its fur and to the condition of the forests. Unrelenting trapping and deforestation during the 1700s and 1800s reduced both the beaver population and their source of food and building materials, until by the mid-1800s the animals were all but extinct in the Northeast. Today, of course, their fortunes have revived, hand in hand with trapping regulations and reforestation.

Beavers often are active at peatlands where there is flowing or ponded waters, especially—but not necessarily—if trees and tall shrubs grow nearby. Consequently, their lodges and dams can be found far more frequently at fens than bogs (where they are restricted to laggs and outflow streams).

But in some special cases beavers are peatland animals to an extent we may not generally realize—in fact, they can even make peatlands. Their damming of waterflows causes paludification, and under the right conditions this leads to the development of peatland. But a contemporary peatland made by beavers may be only the most recent chapter in a long history of development, for the site may well have passed through many centuries or millennia as other wetland types (perhaps an earlier peatland), stream course, or even upland before this most recent episode.

We know that beavers have a long association with some peatlands. Scientists probing a Connecticut bog came across a perfectly intact beaver lodge buried under five feet of peat! Apparently the bog simply overgrew the lodge, long since abandoned by the beavers.

The mobility of the large mammals discussed thus far means that in most cases peatlands constitute only a fraction of their habitats. But many smaller, less obvious mammals spend their entire lives in the bog or fen in which they were born. The fortunes of these little creatures may depend entirely on what is going on within very small segments of those ecosystems.

Some of the same generalizations about small birds hold true for small mammals: none is found exclusively in peatlands, and their presence seems to be due largely to the area's vegetational structure and diversity, as well as to the availability of adequate food. In addition, moisture regimes must be favorable.

The greatest diversity of small mammals occurs where vegetation is thick, plant species are numerous, and the vegetation includes many life forms (e.g., mosses, herbs, shrubs, and trees). The fewest species, not surprisingly, live in nutrient-poor, relatively uniform, open bogs. Forested peatlands may have in their canopies arboreal mammals such as noisy red squirrels (*Tamiasciurus hudsonicus*), nocturnal southern and/or northern flying squirrels (*Glaucomys volans* and *G. sabrinus*), and several species of "woodland" bats (e.g., hoary bat, *Lasiurus cinereus*). Terrestrial species might include the tiny but voracious masked shrew (*Sorex cinereus*) and the herbivorous southern red-backed vole (*Clethrionomys gapperi*). Below the ground surface, if water tables permit, may live fossorial species such as the woodland vole (*Microtus pinetorum*) and, deeper, the burrowing star-nosed mole (*Condylura cristata*) with its almost unbelievable nose, whose pink appendages aid it in discovering invertebrate prey in the underground darkness.

In open peatlands small and common field mammals often scamper around peatland hollows and hummocks, where they make their runways and build their nests. With some patience one can usually see scurrying about many northeastern peatlands meadow voles (*Microtus pennsylvanicus*) and northern short-tailed shrews (*Blarina brevicauda*). Voles (and lemmings), incidentally, although superficially looking like mice, can usually be distinguished by their shorter tails and legs, smaller ears and eyes, and stouter bodies. Shrews, on the other hand, are smaller than mice and are noted for their constant activity; they almost incessantly dash about in search of food (some eat over three times their weight per day!). The meadow vole and short-tailed shrew are good representatives of the two main kinds—from a feeding standpoint—of small mammals living in open bogs and fens: herbivores (mice, voles, and lemmings) that eat the seeds, stems, and leaves of herbaceous plants and carnivores (shrews, moles, and bats) that prey on insects, other invertebrates, or even some small vertebrates.

A few small mammals are so closely associated with peatlands that they are almost, though not quite, peatland endemics. This is especially true in the southern parts of the Northeast, where peatlands provide the majority of suitable habitat. Of particular interest are the volelike northern and southern bog lemmings (*Synaptomys borealis* and *S. cooperi*).

Lemmings, for many of us, belong to fables. We carry the images of stories told in our childhood of thousands of creatures dashing headlong over cliffs on their suicidal mission to the sea. In actuality they seldom, if ever, make mass migrations or suicide trips. In the Northeast these small, secretive, brownish-gray rodents zip around in low vegetation. Active along their runways and in their tunnels among the vegetation, they leave two telltale signs: sedge stems clipped about an inch long and heaped like miniature log piles near their

Fig. 60 *Southern bog lemming (Synaptomys cooperi)*. The southern bog lemming occurs farther south and more commonly in the Northeast than its near relative, the northern bog lemming (*Synaptomys borealis*). It is rarely seen because it is active within obscured runways, burrows, and tunnels in the peatland vegetation. Lemmings cut grass and sedge stems and herbs, making small hay piles along their paths.

travel lanes and bright green droppings (often at special manuring spots), by-products of diets heavy in herbs.

The southern bog lemming is the more common of the two species in the Northeast. Its range spans the entire area, extending south to Virginia. The northern bog lemming, different in subtle physical ways, becomes more frequent northward into Canada. In most of our region it occurs as a disjunct subspecies, recorded rarely in northern New Hampshire and Maine.

We should stop and pause. The fauna of peatlands is part and parcel of those systems. We here have looked at only some of the possibilities, perhaps the more obvious or, for us, the more dramatic. But each peatland has its own complement of creatures—minks or muskrats, solitary sandpipers or short-eared owls, redbelly snakes or spring salamanders, even insects, spiders, and snails. A hand lens and microscope would reveal even more enthralling sights. Together with plants and the many diverse environmental factors, the animals of the peatlands form a total fabric, meticulously spun and woven for ages.

15 | VAULTS OF HISTORY

So far, we have explored primarily the surface of peatlands, its plants, animals, hydrology, and morphology. The living surface is but one result of hundreds or thousands of years of life piled on itself, for beneath lie preserved in the organic remains the record of the past. We have seen only the skin of a much more substantial body. To consider properly the whole peatland, we must examine the long history of development in a library of peat. The stories are long and often complex, but with increasing precision they are being deciphered by trained readers. More than a record of site development, peatlands are, to quote the noted British scientist Godwin (1981), "archives of history," for they provide a window by which we may view earth's environment over great rolls of time.

Peat is the generic term for poorly or incompletely decomposed plant and animal remains naturally formed under conditions of water saturation. It is, in essence, organic matter kept relatively intact in oxygen-depleted environments, usually lying in its place of origin.

Water itself is the main reason that peat forms in the first place. Water retains oxygen poorly, remaining oxygenated only when periodically mixed with air—say, in a splashing brook or when winds produce whitecaps on a lake. Thus, deep or quiet waters become quite anaerobic. We know of

many items well preserved just because they have been submerged in water: wooden ships on ocean bottoms, logs deeply
sunk in lakes. In peats the oxygen-poor internal waters account for most of the "fossilizing," since peats, regardless
of their origin and degree of decomposition, contain about
eighty to ninety-five percent water.

Because the water table in peatlands is high, the rate of decomposition slows drastically from the aerated surface to the
saturated zone just below it, and it keeps decreasing slowly
but steadily down through the entire deposit. Near the bottom, where the peats are usually the most highly humidified,
compacted, and without much oxygen at all, decaying may
well be close to nil.

When pulled apart and analyzed by hand, peat may reveal
entire plant and animal parts—whole plants, seeds, pollen,
leaves, twigs, insect exoskeletons, bones, and so forth—some
of them recognizable by species. Amid these larger remains
and in more decomposed peats (where whole organs are in-
distinguishable) there reside a great variety of minute particles discernible by microscopes, magnetometers, spectro-
photometers, and other sophisticated instruments. Some are
inorganic, such as airborne silt or clay, radioactive fallout,
magnetic particles, and heavy metals. Although most of the
microscopic peat is derived from the peatland organisms, it is
the atmospheric or stream-borne deposits from other environments that permit peatlands to be the repository for information from landscapes far beyond their own boundaries.

The degree of decomposition of peats is one of the main
features by which they often are classified. "Sapric" peats are
the most decomposed, hence usually older and deeper. (The
correlation between peat depth and type or degree of decomposition has many exceptions, which depend on the developmental history of the site.) "Fibric" peats contain many recognizable plant parts and are little decomposed, thus typically
are younger and nearer the surface. Intermediates are called

"hemic." Sapric peats are mushy (sometimes described as being like porridge), and when squeezed, peat oozes from between one's fingers. Plant remains cannot be identified with the naked eye in sapric peats. Fibric peats, on the other hand, retain much of the identity of their precursor plants, and squeezed-out water is clear or nearly so. The "soapy" feel of many sapric peats comes from high concentrations of humic acids accumulated during the decomposition process; the acids are gelatinous colloids that cause the frothiness often seen in streams where organic soils are plentiful.

Besides its great capacity to hold water, peat (aided by its trapped water) also has the ability to insulate. Although summer temperatures at an open peatland surface are equivalent to those of similarly vegetated surrounding lands, just below the surface they drop sharply. Water there remains cool throughout the summer. In fact, at high elevations or in areas with a particularly short growing season, the peats may protect ice pockets from the heat, and ice lenses may persist well into the warmer months. During a spring or early summer walk on the bogs of down east coastal Maine or the peatlands of northern New England alpines, probe with a stick (especially in hummocks) to locate lingering ice.

Peat has other properties that make it an unusual medium for supporting ecosystems. Almost fifty percent of a typical peat consists of humic acids; these acids give the peatland waters much of their acidity and dark color and help preserve organic material. Approximately fifteen percent of peat is bitumen, a substance composed of waxes, paraffins, and resins that can become a prominent ingredient in coal. Given enough time, compaction, and chemical rearrangement, peats will change into lignite, then soft coal (bituminous), then hard coal (anthracite). The immense coal deposits of earth were once, millions of years ago, vast peatlands.

Peat formation and accumulation, besides changing throughout the profile of a peatland, vary with place, climate, and geologic period. Deposition can be incredibly slow: it may take as much as 100 years to build *one* inch of peat! With many of the Northeast's peatlands (0 to 20 feet deep—a few as deep as 45 feet—and 8,000 to 10,000 years old) we can speculate that below the zone of active decomposition, every 1 to 3 inches of peat represents about a century or so of accumulation.

The rates of deposition can be agonizingly slow. Researchers have calculated that the upper 18 to 20 inches of peat in the hummocks of ombrotrophic bogs represent anywhere from 275 to 1,645 years of activity; in the hollows the span is from 325 to 2,190 years. The upper active zone is where the decomposition rates are the greatest; the deeper levels, which are waterlogged, without ties to the living surface and ill supplied with oxygen, progress at rates akin to geologic processes.

The first deciphering of peat profiles was done with the help of billions of microscopic pollen grains shed by nearby and distant plants. Whether deposited directly or transported in by air or water, the pollen, year after year, lays a light blanket that will be preserved centuries on to millennia. Pollen of many species have hard, virtually indestructible chitinlike shells around the softer organic contents; thus, they last almost indefinitely in peats. What makes the grains so valuable for analysis is that the pollen of different genera and species differ in size, shape, and patterning. These traits can be distinguished by trained technicians when viewed microscopically.

Obtaining a sample of a peat profile is itself a straightforward procedure. Several kinds of special coring tools can extract from any desired depth. The contents are examined grossly in the field, then taken to a laboratory for pollen extraction. If enough samples are taken through the entire depth of peat, at regular intervals, interpretation of pollen assemblages will yield a general picture of what was growing in or near the peatland at various times in its history.

Interpretation of peat stratigraphies, however, is a laborious and often confusing job. To begin with, different species of

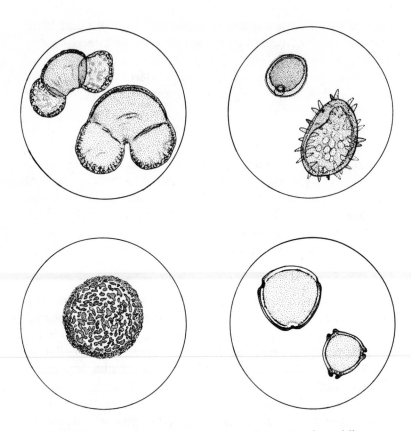

Fig. 61 *Pollen grains*. Under a microscope pollen grains show differences that allow scientists to distinguish among plants. Pollen of several conifers have inflated sacs, assisting their airborne dissemination. At *top left* the larger grain is that of black spruce (*Picea mariana*), the smaller of pine (*Pinus* spp.). Grass (family Poaceae) pollen has few distinguishing features (the smooth-coated grain, *top right*). Pollen with spikes like those of pond lily (*Nuphar* spp., *top right*) are sometimes causes of hay fever. Some pollen has highly sculptured surfaces, as do the roughened grains of eastern hemlock (*Tsuga canadensis, lower left*). Many pollen grains are subtriangular in surface view: at *lower right* the smaller grain is paper birch (*Betula papyrifera*) and the larger is bitternut hickory (*Carya cordiformis*).

Fig. 62 *Peat coring*. In recent years many specialized corers have been devised to extract samples from various depths in different peatland conditions. This Macaulay corer (originally designed in the USSR) is used widely in peat research. In this sample from the bottom of a Vermont bog, one can readily see the transition from older lake clays (light sediments) to the initial bog peats (dark). *(Photo by Stephan Syz.)*

plants shed pollen in different quantities. White pine, for example, produces a great deal of pollen, whereas balsam fir puts forth comparatively little: one white pine can produce more pollen than hundreds of firs. So simply measuring the quantity of pollen that a species has shed in a peatland does not give an accurate picture of the abundance of that species. Furthermore, because most pollen is light and can be carried great distances by the wind, its presence may not give an accurate account of what was growing in the vicinity of the peatland. One researcher found, for example, that spruce and fir pollen accounted for twenty percent of all pollen in some Canadian arctic peatlands, but those peatlands were 250 miles north of where spruce and fir occurred.

Other problems hamper efforts to determine the relative

ages of the peat layers from which the pollen was taken. As older peats are buried under the younger, they are increasingly compressed, compacted, and distorted. This distortion, together with the fact that peats accumulate faster in some parts of the peatland than in others, means that the same level in two different locations of the same peatland may represent two very different periods. Burrowing animals, in addition, may redistribute peat and pollen. The sophisticated technique of radiocarbon dating permits fairly precise estimates of the actual ages of peats, but this process is both costly and time-consuming and likely to be out of reach for many field biologists.

Radiocarbon dating was used, for example, at Shelburne Pond and peatlands in northwestern Vermont in conjunction with analysis of pollen, preserved diatoms (microscopic algae with delicately sculptured, grasslike encasements), and gross peat type to determine the long-term history of the 440-acre pond and its environs. Figure 63 displays the pollen abundance for selected species in peats beneath the sedge-sphagnum peatland. Figure 64 summarizes the various data and their interpretation. The following account refers directly to these two figures.

After glacial ice left the basin some 12,500 years ago, an extensive regional ice-dammed lake (Lake Vermont) deposited clays widely over the landscape. When that lake drained about 11,700 years ago, a local lake (in the basin now occupied by the pond and wetlands) continued to fill with clay—clay containing pollen from the first neighboring vegetation, a mix of tundra with spruce, fir, and pine woodlands (zone A). As time passed, spruce and fir declined, leaving pine to dominate. During this period (B, about 9,500 to 7,400 years ago) only a meager amount of algae grew in the lake.

There followed a somewhat warmer and drier period (generally called the hypsithermal interval) when hemlock increased and northward-migrating hardwoods such as oak and birch reached the watershed. At the height of the hypsither-

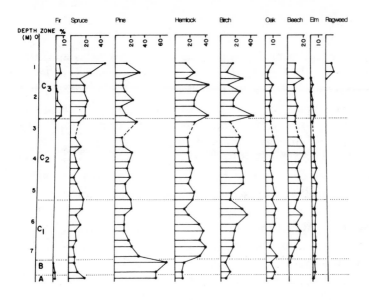

Fir Spruce Pine Hemlock Birch Oak Beech Elm Ragweed

DEPTH ZONE (M)

Figs. 63 and 64 *Pollen profile.* This chart shows relative amounts of pollen of different genera of trees (plus one herb) at various depths in Shelburne Pond Bog, Vermont. In this type of analysis the cumulative widths of the column at any given depth horizon total 100%. Thus, when looking at the entire column of pine, for example, one can say that at the depth of about 4 meters (approximately 12 feet) pine pollen comprised about 20% of all pollen recorded. But one cannot determine if that is also the period with the greatest influx of pine pollen per unit time. With the help of radiocarbon or other dating techniques, depth can be correlated with age. See the text for interpretation of this pollen profile.

Time	SHELBURNE POND				Regional climate (a)
Thousands of years before present	Wetland core			Basin physiography	
	Sed. type	Pollen	Diatoms		Warm / Cool
0					
1	Sphag. peat	C₃		POND	
2	Sedge peat		D IV	BAYS FILL	
3		C₂	D III		Cooling
4	Sedge and		D II		Thermal maximum
5	algae			LOCAL	Hypsithermal period
6	gyttja	C₁	D I	LAKE	
7					
8	Clay gyttja	B			
9					
10	Clay	A			Warming
11	?				
12	?			REGIONAL LAKE	
13	(Ice)			GLACIATED	

(a) After Bradstreet and Davis, 1975

mal interval during C_1 (some 7,400 to 4,300 years ago) the free-floating diatoms (D I) grew abundantly enough to leave interpretable remains. By the latter part of this period algal populations increased, the shallower bays began to accumulate significant organic sediments, and bottom-loving (benthic) diatoms increased markedly (D II).

As hemlocks declined in the waning hypsithermal, a mixed hardwood/softwood forest formed (C_2, 4,300 to 1,600 years ago), bays filled with sediments and became peatlands, and benthic diatoms could thrive (D III) in the now much shallower lake. When the peatland vegetation closed over the algal sediments, the diatoms ceased to be a major contributor to the peats (D IV).

Currently the pond continues to fill with organic sediments, the pollen (C_3) reflecting the transition from a native landscape to one colonized by Europeans, whose land clearing is marked in other profiles by significant amounts of ragweed (*Ambrosia* species) pollen. Two of the peatlands are transforming from sedge fens to sphagnum-heath bogs; projections estimate that the 440 acres of open water will close with peatland vegetation within the next few hundred years.

But the deep "pages" of a peatland furnish an even more comprehensive history than that of past plant communities and climates. Charcoal layers tell us that most peatlands have fires from time to time and that some burn regularly at intervals of 200 years or so. Minute amounts of glassy ash from distant volcanoes record eruptions and the path of the ash-carrying winds. Plantain (*Plantago* species) and ragweed (*Ambrosia* species) pollen in quantity mark the advent of colonial clearing and agriculture. Mastodon and mammoth bones help to demonstrate that these extinct ice age boreal giants once roamed the Northeast.

Disturbing indications of more current ways of life also are being recorded. Lead, until recently an ingredient in automotive fuels, appears in the peats laid down since the invention of the automobile. In the layers deposited between 1920 and 1950, and again between 1950 and 1970, lead shows a dramatic increase, corresponding to the surges in use of cars and trucks in this country. In recent deposits magnetometer measurements trace the pathway of aerial industrial pollutants. Peatlands also record the recent resurgence of DDT in our environments; although now prohibited from use in the United States, much is exported to Mexico and Latin America from where it is carried by winds back to this country (some evidence suggests surreptitious uses in the United States as well). Other products of our industrial age also are trapped in the layers: PCBs, radioactive fallout, heavy metals, and other pollutants are present in high amounts in recently deposited peats, revealing the abundance of such substances in the atmosphere in the last forty years and marking a new course in our history.

The plants and animals that greet us in peatlands are but the latest mantle of a continuous historical legacy millennia old. Our plans to dredge or mine must recognize that not only will a particular living landscape be lost but so too will a library of the past. Our paths should be stepped gently, in respect for the living surface, of course, but in reverence as well for the heritage beneath.

16 | HUMAN USES OF PEATLANDS

Little detailed information survives about how primitive American cultures used peatlands. What evidence we have indicates that, in keeping with their ways of life, the native peoples of the Northeast made consistent but not intensive use of peatland plants, mostly for medicinal or magical purposes (Table 7). But in contrast with what has followed since European colonization, Amerindians indeed made slight use of peatlands.

In the same region today, if people were asked what kinds of goods we get from peatlands, most would probably answer "blueberries" or "cranberries"; others might suggest "peat moss." These responses, although generally correct, do not paint a complete picture for the Northeast. We do indeed raise a lot of cranberries and some blueberries, but we also raise a few other crops, notably vegetables (primarily in New York). A small number of mining or harvesting corporations (mostly in Maine) extract peat for horticultural uses. Nonetheless, most peat moss sold in the Northeast comes from New Brunswick, Quebec, and North Carolina, and the majority of peat-grown vegetables we eat comes from the Midwest and Atlantic Coastal Plain. Most peatland use in the

TABLE 7 *Indian Uses of Some Bog Plants*

Plant	Use
Sphagnum mosses (*Sphagnum* spp.)	stuffing for pillows and mattresses; "diapers"
Tamarack (*Larix laricina*)	treatment for scurvy, bronchitis, infections; rope, twine, and caulking (for canoes, etc.); beer
Pitch pine (*Pinus rigida*)	treatment for tuberculosis
Cotton grasses (*Eriophorum* spp.)	food
Bog willow (*Salix pedicellaris*)	treatment for stomachache, pain; astringent; tobacco
Sweet gale (*Myrica gale*)	preservative in blueberry pails; burned as smudge to repel mosquitoes; reddish-brown dye; charm against snakes
Gray birch (*Betula populifolia*)	treatment for bruises, wounds, cuts, scalds; pain during childbirth and menstruation; pipe stems
Pitcher plant (*Sarracenia purpurea*)	drinking cup; toy ("frog leggings"); cup for berries; treatment for smallpox, kidney and lung ailments; sorcery
Bog rosemary (*Andromeda glaucophylla*)	tea
Leatherleaf (*Chamaedaphne calyculata*)	treatment for fever, inflammation; tea
Labrador tea (*Ledum groenlandicum*)	treatment for ulcers, burns, fever; blood cleanser; expectorant; tonic; tea; brown dye
Blueberry (*Vaccinium* spp.)	food; treatment for insanity, pleurisy, pain during childbirth; blood cleanser
Cranberry (*Vaccinium* spp.)	food; treatment for nausea
Bog goldenrod (*Solidago uliginosa*)	treatment for boils
White bog orchis (*Platanthera dilatata*)	love charm
Reindeer moss lichen (*Cladonia rangiferina*)	wash for babies
Sheep laurel (*Kalmia augustifolia*)	treatment for headache, backache, colds
Creeping snowberry (*Gaultheria hispidula*)	tea

SOURCES: Adapted from Densmore (1974), Weiner (1980), and Kovacs (1979).

Northeast has been by conversion (draining, dredging, filling, or flooding) into non-peat agricultural lands, development land, or reservoir bottoms. Most extant peatlands—and there are many, especially in the northern states—have been left alone and are little altered.

This is certainly not the picture, however, in many other countries where peatlands are prevalent. Most prominently in the Soviet Union, Finland, Scandinavia, Germany, Great Britain, and Ireland, people have used peatlands over hundreds and even thousands of years for fuel, crop and pasture lands, and forest products. In recent decades technology imaginatively has created many new uses and greatly increased extraction and conversion rates. This technology is aggressively marketed worldwide by all the major producing nations, especially Finland, Ireland, and the Soviet Union.

Various countries harvest naturally growing products, including fruits (mostly cranberries, blueberries, and cloudberries), trees for lumber and pulp, and hay for livestock. Peatlands are prepared *in situ* for agriculture: foods (root crops, greens, berries), forestry plantations, hay and grain crops, pasturage, and most recently, biomass production for energy (using species of hybrids of willow, alder, poplar, and cattail). Peat is mined for fuel (for use in homes, electrical power plants, and industry), for chemical extraction, for processing into, for example, building materials, insulation, plant pots and seed starters, and for horticultural additives (the familiar "peat moss" sold at garden stores).

Today, as for most of this century, the principal uses of peatlands in countries other than the Soviet Union and Ireland is for forest products, agriculture, and horticultural additives. Several properties make peat useful when growing plants. It is, of course, highly organic and acts like a mulch when applied to a clayey or dry soil, improving water retention and mobility and increasing tillability. When applied to nonsaturated soils, peat not only makes heavy soils lighter and more friable but also releases, as it decays, nutrients such as nitrogen and phosphorus. When added to waterlogged or fine-textured soils such as clay, the peat provides much needed ventilation: air can penetrate more easily. But as an additive to dry soils or in droughty areas, the great water-holding capacity of peat, especially sphagnum moss peat, retains soil moisture. (Interestingly, research is underway at Pennsylvania State University to use live sphagnum moss as a ground cover for old strip-mine sites, since it can tolerate the highly acid condition of such places.)

In the Northeast bagged horticultural peat is produced primarily (and perhaps exclusively) at a few raised bogs in Maine, which has an average export of 5,000 tons a year (in contrast with neighboring New Brunswick's annual sales of 120,000 tons). Raised bogs provide sphagnum-rich peat, ideal for most soil conditioning needs, and, by virtue of being raised, they are much easier to drain than peatlands confined to basins. Mining is by the so-called milled-peat method. After ditching and allowing about five years for draining, the vegetation is scraped aside until the water table is neared, or more commonly, it is rototilled into the surface peats. Then after a good day or two of sunshine, drying air, and occasional light tilling, the top ¼ to ½ inch of drier peat is vacuumed up and hauled to storage piles for subsequent screening, bagging, and marketing. Several smaller operations scattered around the Northeast dredge or scoop wet basin peats and deliver them for soil additives by dump trucks. These more highly decomposed peats contain more sedge and/or wood and less sphagnum; at one southern Maine site the peats were chosen specifically for their exceptionally high nutrient content.

Many northeastern peatlands have been cultivated for agricultural use, especially in New York and Massachusetts: rather than bringing peat to the crops, people plant crops in the peatlands! Preparing peatlands for such purposes involves considerable effort and expense. Usually the site must first be drained (the extent depends on the crop), heavily fertilized, and groomed for the particular species to be planted. But

Fig. 65 *Peat mining*. This "Martian Bigfoot" vacuums up peat from Denbo Heath, a large peatland in eastern Maine, eventually to be sold as "peat moss" for horticulture. Vacuuming is part of the "milled peat" mining process. After water tables are lowered and vegetation cleared from a raised bog, harrows or tillers fluff up the surface. After several hours of favorable drying weather, the drier upper 1/4 to 1/2 inch of peat is vacuumed. The dried peat is carried to large storage piles for subsequent bagging for sale— or as in Ireland and the Soviet Union, taken to power plants for burning.

once this preparation is accomplished, peatland can produce quite a garden. Cold-weather species do best, but under good management others can do well. Grown in the Northeast are vegetables such as corn, soybeans, carrots, lettuce, onions, radishes, and parsnips, fruits such as strawberries, raspberries, and cranberries, and a variety of flowers.

Of all northeastern peatland crops, the most famous and commercially valuable is the cranberry (*Vaccinium macrocarpon*). The tart red berry is at the heart of an extremely large industry centered in the small state of Massachusetts. In

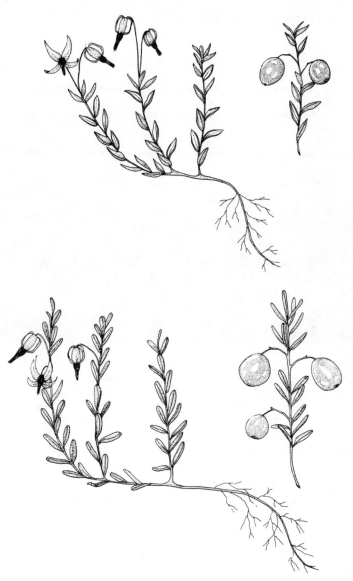

Fig. 66 *Cranberries* (*Vaccinium oxycoccos* and *V. macrocarpon*). In the Northeast two cranberry species commonly occur in peatlands. Of the two species, small cranberry (*V. oxycoccos*) (*top*) has a more wiry stem, with smaller and more pointed leaves, and is more prostrate. It thrives in raised bogs and other nutrient-poor peatlands. Large cranberry (*V. macrocarpon*) (*bottom*) is a plant of more nutrient-enriched sites and is the species grown for commercial harvest.

Figs. 67 and 68 *Cranberry harvesting.* Two ways of gathering cranberries in the fall in Massachusetts. Dry harvest, the older way (*top*), is good for obtaining undamaged whole berries; wet harvest (*right*) is more efficient, and most of its yield goes into juices and sauces. Wet harvesting begins with flooding of the bog. In one method machine beaters loosen the ripe berries from the plants, and the wind blows the floating berries to one side, where machines with conveyor belts pick them up and put them in trucks. *(Photos courtesy of U.S. Department of Agriculture.)*

fact, well over half of the world's trade in cranberries comes from Massachusetts, mostly from the southeastern counties, including Cape Cod and Nantucket Island. Other states, notably Wisconsin and New Jersey, also have commercial cranberry enterprises, but Massachusetts currently leads in production. In addition to natural peatlands, many "bogs" are now artificially created (unfortunately, often at the expense of other wetlands) or highly modified to encourage large yields and to accommodate harvesting equipment.

Though not normally considered a crop, trees are planted

in peatlands in many regions of the world, notably Finland (over 13 million acres), USSR (over 11 million acres), and Scotland and Ireland (well over 750,000 acres). Significant plantings are proposed for areas of Canada's maritime provinces and Maine. However, in the Northeast forest products are, and have been, natural in origin. Black spruce (*Picea mariana*), which can be turned into high-quality paper, is an important species to the pulp-and-paper industry. Tamarack (*Larix laricina*), less valuable for paper, is used for products that need to be rot-resistant, such as fence posts, railroad ties, and sills for houses. Since the earliest colonial times northern white cedar (*Thuja occidentalis*) and Atlantic white cedar (*Chamaecyparis thyoides*) have been major forest product species. From cedars have been made fence posts and railings, railway ties, shingles, siding, ship boards, chests, telephone poles, and a wide variety of other items. (Cedars continue to be cut today, wherever they grow, except at a few preserves and scattered sites in remote locations, especially northern Maine.) Atlantic white cedar apparently was a material used for early pipe organs—the story is that the inventor got the idea from listening to the beautiful resonance of rain falling on cedar shingles.

Various peatland plants have been used locally for foods and home remedies, often drawing on older European traditions. Of course, there are blueberries and other berries picked for pies, preserves, and jellies. Cloudberry (*Rubus chamaemorus*), long harvested in Europe, Scandinavia, and Russia, has a fruit considered delicious by many; it is used for pastries and preserves in eastern coastal Maine. The large soft fruit tastes like and has a texture similar to baked apples—hence its other common name, "baked-apple berry." In Fennoscandia the fruits are further processed as an expensive liqueur. The British and colonial Americans put leaves of sweet gale (*Myrica gale*) in pillows and blankets to scent them and to repel fleas. Labrador tea (*Ledum groenlandicum*) occasionally has been picked to make a strong tea. And no doubt other plants provide leaves, fruits, bark, or flowers for folk medicines here and there about the Northeast.

Peatlands have also provided somewhat unexpected substances for human use. By a complicated series of reactions within the peat moss (and in pond and river bottoms, in some instances), certain bacteria in the absence of oxygen can precipitate ferrous oxides and hydroxides in such large quantities as to make consolidated deposits of a low-grade iron ore called "bog iron." Until supplanted by the discovery of richer and vaster ore deposits in Pennsylvania and Minnesota, this ore was mined and smelted in many places in North America and abroad. The New Jersey Pine Barrens had a substantial bog iron industry between 1765 and 1865, with some thirty blast furnaces and forges processing the ore extracted from the peatlands and streambeds there. Minor mining operations served local needs at scattered locations elsewhere throughout the Northeast in the nineteenth century.

Basin-fill peatlands in the Northeast, where there are limestone bedrock or limy tills, often have precipitates of calcium carbonate mixed with clay, called marl. From time to time peatlands have been drained or dredged to retrieve marl for use as a lime fertilizer for acidic croplands or as an ingredient in mortars.

In addition to growing crops, improving soils, and serving as fuels, peat is useful to humans in other ways. The efficient insulating properties of sphagnum mosses have been recognized for centuries. Rural Scots filled hollow walls of their stone "black houses" with it; Russians incorporated fibrous peats into home insulations; Norwegians have used it in railroad beds to lessen frost heaving.

In the Soviet Union particularly, peat is used in an astonishing number of other ways. It is blended with wood fibers and glue to make peatwood, similar to plywood; it is prepared with other substances to make "peat cork" and "peat foam"; and it is mixed with cement to make light but tough "peatcrete," similar to concrete blocks. The Soviets also have extracted from it an array of important chemical products: ethyl alcohol, waxes (used in lubricants, polishes, and candles), and resins (as synthetic steroids in medicine); used it as an organic slurry for culturing yeasts to produce alcohol for human consumption or high-protein animal feed; and transformed it into "peat coke" for use in electronics, "peat tar" for pesticides and wood preservatives, and a kind of gunpowder and fuses for explosives. None of these uses is now employed in the Northeast, but several have been considered, at least in a general way, by industry and by a few state and federal agencies.

In several countries (notably Finland) peat and peatlands are put into rather unglamorous service for sewage treatment, since the peats remove phosphorus and nitrogen through natural processes. Untreated wastes may be deposited directly into a peatland and the living system allowed to deal with them as it will. This usually results in increased plant vigor and a change to more fen or eutrophic communities. It can be employed only where peatland alteration can be tolerated and where the volumes of effluent is minuscule compared to the annual water budget of the peatland. In another method, wastes are deposited into a drained and ditched peatland, where they can be assimilated more quickly and thoroughly. Or they may be sprayed into a peatland that has been remade into a peat-and-sand filtering system. Finally, in many instances the peats are excavated and brought to the filtration site, where either a filtration system is constructed or an artificial peatland is created. In the latter case, grass is usually sown on the peatland; as the grass grows, it takes up much of the nitrogen, phosphorus, and other effluent nutrients. The mown grass then can be used for mulch or composted for eventual fertilizer. (Dried sphagnum peat added to human wastes aids composting; the extra carbon raises the carbon-nitrogen ratio. In the Northeast some families employ this method in home-composting toilets; the U.S. Forest Service also uses it at some back-country outhouses.)

These systems are quite new or still experimental in the United States. Several small, private peat treatment systems are in operation in the Northeast. During the last few years several proposals to use bogs *in situ* have been tendered (at least in Maine and Vermont). The uncertainties of the ultimate changes in the peatland and the length of time before a site can absorb no more nitrogen (and other nutrients, heavy metals, and so forth) have restrained development.

In their natural or very slightly altered state, peatlands provide a variety of benefits, from hydrologic buffers to scenic vistas. They are valued as critical habitats for game and nongame species, as well as for rare and endangered species. They sometimes serve important roles as biotic filters in watersheds (removing suspended materials and acting as sinks for toxic materials and heavy metals), as mediators of potential floods, and as places for passive recreation and aesthetic appreciation.

The larger, wetter, richer fens provide hunting, trapping, fishing, and occasionally boating, which, when combined with their water-purifying ability, often result in their protection and management by state or federal agencies. Recently throughout the Northeast there has been a dramatic increase in the use of peatlands—especially bogs, nutrient-poor fens, and sites rich in orchids or carnivorous plants—for nature study, scientific research, and aesthetic enrichment. This in turn has led to the creation of preserves to maintain habitats for species, communities, and/or ecosystems. Especially near larger residential areas, the few more or less natural peatlands receive tens or hundreds, and in a few cases, over a thousand visitors a year. Some sites have boardwalks, but most show signs of trampling, often as heavily abused vegetation and deep trail scars. Several peatlands are much researched, being revisited on a regular basis by scientific parties. In contrast to these sometimes disruptive uses, protecting a peatland as a preserve often means severely limiting visitors.

TABLE 8 *Comparison of Fuel Peat with Other Fuels*

	Heavy Fuel Oil	Coal (Bituminous)	Coal (Lignite)	Peat	Wood
Carbon (%)	86	82	70	55	49
Nitrogen (%)	0.25	1.0	1.5	1.8	1.4
Sulfur (%)	2.4	2.0	2.0	0.2	—
Ash Content (%)	0.3	4–10	6–10	2–10	0.5
Volatiles (%)	—	10–50	50–60	60–70	75–85
Operational Moisture (%)	0.1	3–8	40–60	40–60	30–55
Effective Heat Value (kcal/g) at the Lowest Operational Moisture Content (i.e., well-dried)	9950	7100	2900	2750	3000

Adapted from New York State Energy Research and Development Authority, 1982. Most figures represent averages of ranges. Values may vary somewhat depending on source or type of fuel in each category. For example, hickory wood will yield high heat values; pine gives lower.

Looming over agricultural, horticultural, forestry, and other peatland uses is the demand for fuel peat for use in heating and electrical generation. In countries with established fuel peat industries use levels are rising dramatically, and many new countries are seeking to develop fuel peat systems. The economic and logistic problems caused by the need to dry and transport the peat are slowly being solved—technological methods are being devised to convert peat to liquid and gaseous forms, thereby allowing safer and more efficient transport.

In northern Europe (now primarily in parts of Scotland and Ireland), peat use as a domestic fuel is a time-honored tradition. It is cut annually in the spring, dried in stacks, and burned in stoves and fireplaces (and formerly on open hearths). The practice is ancient: the mummified "bog people" of Denmark were found in old trenches where peat was cut some two to three millennia ago. Peat likely became more important as a fuel after the great deforestation of Europe, Great Britain, and southern Scandinavia. With the increased use of coal, then oil, peat as a fuel declined until the 1950s. National programs of peat development in Ireland, Finland, and the Soviet Union in the postwar decades have resulted in the con-

TABLE 9 *Peat Extraction, by Country*

Country	Percentage of World Harvest
USSR	92
Ireland	3
Finland	2
West Germany	1
China	1
All Others (including USA, Canada, UK, Poland, Sweden, Norway, and Malaysia)	1

SOURCE: Adapted from Kivinen and Pakarinen (1981)

version of hundreds of thousands of acres of peatland, through mining, drying, and burning, into peat for electricity and district heating.

Large-scale industrial use of peat began in the early decades of this century. Russia introduced its first peat-burning generating plant in 1914. Now the Soviet Union has more than seventy-five, consuming over 200 million tons of peat annually (yet this is only two percent of the nation's energy requirements!). Some of the plants are capable of producing 630 megawatts of energy. Ireland, with six peat-fired plants, generates about one-third of all its electricity with peat, burning some five million tons per year. Finland anticipates tripling its use of fuel peat in the near future.

In the search for domestic sources of energy as an alternative to oil sparked by the oil crisis of 1972, industry and governments in Canada and the United States have added this continent's peatlands to the list of potential energy sources. Interestingly, shortages of wood, coal, and oil during World Wars I and II prompted peatland inventories in this country, Maine being one focal point because of its good-sized raised bogs.

North America has immense deposits of peat vulnerable to exploitation, enough to make the prospects attractive and promising. Various feasibility studies considering the potential for large-scale fuel peat production, funded by governmental departments of energy and by energy-based corporations, have been initiated in several Canadian provinces and at least in the states of Minnesota, Alaska, North Carolina, and in the Northeast, Maine and New York. The U.S. Department of Energy has invested millions of dollars both in assisting these studies and in exploring other possibilities on its own.

In the Northeast, resource surveys that include inventories of peat deposits, site surveying, and general economic assessments have been undertaken in Maine and New York. Maine is where the potential for fuel peat development has become sufficiently imminent to create a flurry of great interest and heightened concern. A current proposal would have several thousands of acres of streamside raised bog in central Maine converted to energy, perhaps by a method not requiring drying of the peat.

In Minnesota a gas corporation proposed to lease 250,000 acres of state-owned peatlands for the purpose of converting mined peat into methanol gas to be piped to outlying residential areas. There governmental agencies, industry, and environmental groups—through research, due process, and lobbying—have created much insight into the economic and environmental consequences of major fuel peat production. In Maine fuel peat proposals in the last eight to ten years have drawn considerable response from concerned citizens, environmental organizations, and planning agencies, resulting in burgeoning meetings, conferences, policies, and biological studies about the state's peatlands.

The potential impacts of large-scale fuel peat production are staggering. A power plant of only 100 megawatts would consume in twenty years about 10,000 acres of peatland to a depth of five feet. In Ireland an electrical generating plant built in the 1950s will be shut down in 1985 or 1986, since there will exist no more peatlands sufficiently close to supply it. During final months of the plant's operation the last of one of Ireland's four major peatland types will become extinct as it is mined and trucked away to burn. Not only does mining directly alter indigenous plants and animals, but it affects local hydrologies and downstream water flow and quality. Local climatic changes have been postulated. And from an ethical point of view, the cashing-in of landscapes that are essentially wild and natural raises many questions of human values and perceptions.

An increasing number of sites are being left in their natural state, most for hunting, natural history, and science. A lesser number are managed to protect threatened natural features or to preserve aesthetic values. Thus, our often-conflicting desires for attractive natural landscapes, abundant energy, and fertile gardens compete head-on at our larger peatlands. This is a subject to which we shall return in the final chapter.

17 | PRESERVATION OR OBLITERATION?

In our time, in parts of the Northeast and in many countries of the world, peatlands face extinction. The relentless march of overflowing cities and suburbia takes them one by one; governmental policy to supply short-term energy with long-term peats thousands of years old takes them by whole landscapes. International policy spreads worldwide as state-owned industry seeks to, as is declared in one motto, "develop the world's peat resources."

Public interest in the peatlands of the Northeast has had recent awakenings, though even with its distant roots it is still in an infancy. Scientific inquiry is at an all-time high, and our understanding of peatland diversity and ecology has made immense strides during the last ten years. With threats of large-scale mining for fuel peat and the recognition of continued incremental loss of peatlands as populations expand, state agencies and private organizations have instituted numerous local and statewide assessment and protection activities since the mid-1970s. In a few parts of the Northeast issues have transcended the specialists and entered the public domain, the state of Maine being the notable example.

Where, then, does this new awareness lead? What are the

benefits and losses associated with this use or that? What are the best criteria for decision making? The remainder of this chapter seeks to respond, at least in part, to these difficult questions.

Peatlands are valued, and when valued, they are used (whether for mining or for aesthetic vistas). They have values (and therefore uses) because of their *land*ness, their *wet-land*ness, their *peatland*ness, and their *natural*ness. Some uses have been mentioned in earlier chapters, but their numbers are immense. In an address before a conference in 1983 entitled "Protection of the Ecological Value of Maine Peatlands," Ian Worley gave 9 landness, 43 wetlandness, 71 peatlandness, and 135 naturalness values discovered in a preliminary survey (and the listing did not include the several hundred uses for peat employed worldwide). It is no wonder, then, that conflicts arise concerning the appropriate use or management of a particular peatland.

As this book is about the natural features of peatlands, the greatest concern centers on selecting those features worthy of preservation and identifying uses that produce irreparable damage or destroy the irreplaceable. Having discerned these features and uses, what can one do to preserve and protect where warranted?

Finding essentially natural sites in parts of the Northeast (and elsewhere in this and other countries) can be difficult, for civilization has already left deep scars on earth's peatlands. We have drained them for farming; we have crossed them with railroads, canals, roadways, and pipelines; we have put military electronics bases on them; and we have even built towns and cities on them. And we haven't stopped there— we now vacuum them and bag them for sale at supermarkets; we push them aside to make trout ponds; we suck them up to put on gardens; we plant them in lettuce and carrots; we cut each new grown cedar in its youth to fence in cows or fence out strangers; and we propose to burn them by the hundreds

TABLE 10 *Virgin and Protected Peatlands*

Country	Percent of total peat- land area of country that is virgin	Percent of peatlands in country that are protected
Canada	99	?
China	92	0
USSR	91	0.9
USA (Minnesota) *	90	6.7
Norway	83	0.1
Sweden	83	2.1
Czechoslovakia	56	15.9
Ireland	53	1.1
Finland	42	2.0
New Zealand	37	33.0
Great Britain	30	?
Switzerland	21	9.1
West Germany	9	3.1
Poland	6	0.4
Denmark	3	0.004
East Germany	2	1.8

SOURCE: Adapted from Kivinen and Pakarinen (1981), based on data from an International Peat Society survey in 1979 and from Goodwillie (1979).

* The only available data for the USA were for the state of Minnesota. That 90% of the United States' peatlands are virgin takes into account the untouched Alaskan peatlands, which constitute some 75% of the nation's peatlands. In some states there must be less than 25% of peatlands that are little touched. In the Northeast the figure ranges from 95% essentially virgin in Maine to a small, but unknown, percentage in Rhode Island and Connecticut.

and thousands of acres to light our homes and power our televisions.

In use there can be further mistreatment. In Europe, as a result of construction activities on peatlands, "bog bursts" leave great rents of erosion, as potentially devastating flows of highly liquefied peat rush madly downslope. Apparently, too much tinkering with the bog's surface can result in increased pressures on the lower peats, a weakening of the vegetation mat, and finally catastrophic collapse of the whole structure during heavy rains.

Harm and degradation also come in more subtle ways. In the Northeast peatlands have provided many a dumpsite for landowners, towns, and municipalities. Their laggs are inflicted with upland refuse. Unscrupulous and unthinking collectors and flower lovers have locally reduced or eliminated plants fascinating by their unusualness or valued scarcity. Strips of the living vegetation mat are cut, rolled, and shipped to urban centers for window displays in stores.

Ironically, increasing public interest in peatlands as natural areas has brought a new problem—trampling—that kills plants, flattens hummocks, breaks through root mats to underlying peats, and leaves long-lasting trails. Though this damage is unintentional and often incidental, some peatlands are in danger of being "loved to death."

And of course there now looms another potentially revolutionary specter—the exploitation of peat for energy production on a scale hitherto unknown in the Northeast. Our craving for energy pushes us ever closer to the reality of peat-burning power plants, fed by thousands of acres of surrounding wetlands. Maine and New York have peatlands sufficiently large to support the industrial mining of fuel peat. But all regions of the Northeast have smaller peatlands scattered here and there that are possible sources of peat for home heating if a suitable technology can be created that puts peat extraction in the hands of a homeowner. That technology is near at hand; in Europe one can buy or rent an attachment for farm tractors that will extrude a family's annual fuel requirement in an afternoon, a task formerly taking two to three weeks of cutting peat by hand. With similar devices, some sort of roadside peat sump-pump, or other invention of the Yankee ingenuity—the mining of peat may become as personalized as the cutting of firewood by chainsaw. The ramifications could be awesome!

The reason for preserving peatlands in their natural state are many, though few persons would argue that all must remain intact. There issues a will from our society that portions of our natural heritage be protected. At the very least, it would seem, we should retain (a) exemplary sites representative of each peatland type in physiographic regions (such as the unglaciated Pennsylvania highlands) and the individual states; (b) rare peatland types and peatlands that are habitats for rare or endangered species; and (c) peatlands locally valued for education, nature appreciation, landscape diversity, historical significance, and any of a number of other hometown reasons.

Despite the damages and losses already inflicted on the Northeast's peatlands (especially in certain areas), despite actual and potential threats, and despite a growing distaste for the destruction of natural features, with a few notable exceptions there exist few protective shields in law or regulation. Neither the federal government nor any state government has a peatlands policy or law. We do not stand alone in this, for few countries of the world have had the foresight to secure large tracts of peatland for preserves or natural areas prior to rampant development. A noteworthy exception, Finland, has a national program to develop a system of rare and representative peatlands by providing total protection to some areas, partial protection to others, and specific protection of critical features in still others. However, this meritorious and commendable program came into existence only after the vast majority of Finland's peatlands had already been irreversibly altered.

In parts of the Northeast undisturbed peatlands are virtually nonexistent. Yet in large regions, such as much of Maine and mountainous New York, more than ninety percent of the peatlands remain essentially natural. Protection programs thus have the duty in some areas to secure from destruction the last remnants of an ecosystem type, whereas in other areas there is the rare opportunity to select outstanding sites for the preservation of natural values. Speaking of Maine's peatlands (at the conference mentioned previously), Ian Worley stated:

By virtue of . . . their extraordinary diversity and remarkable pristineness, Maine's peatlands provide an excellent array to allow for the protection of a variety of values. . . . In this nation, in no other state save Alaska, does such an opportunity arise to select wisely among a nearly complete, essentially unblemished edition of naturalness. In most states and nations protected peatlands are no more than the unwanted remnants left following decades and centuries of disruption. But in Maine wise decisions can be made *before* wanton destruction yields cast-off tailings.

One survey in the Northeast estimates that less than 0.6 percent of the peatlands have deliberate protection through ownership by a public or private agency or by deed restrictions on the property. Apparently no more than one to two percent enjoy fortuitous security as part of publicly owned parks, forests, or wildlife refuges. These percentages fall far short of what is necessary to preserve examples of the splendid variety of these ecosystems (which contain the last significant virgin landscapes of the Northeast) and the rare elements they contain, safely for the future.

Many kinds of protection can be afforded peatlands, but no greater protection is there than that generated by concerned individuals. One of the earliest champions of bogs in Vermont was Lucy M. Bugbee, who, long before local conservation activism became fashionable, pioneered with missionary zeal the preservation of peatlands. She proselytized throughout the state in schools, civic organizations, hunting and fishing clubs, and anywhere anyone would view her pictures and hear her loving admiration of the bogs, their vegetation, and their flowers. Her contagious enthusiasm and unceasing vigor captured the imagination of many, and even in her eighties she led tours of naturalists and ecologists into favorite bogs to share her fascination with these special places. One of the earliest peatland nature preserves in Vermont now bears her name, for through the assistance of the New England Wildflower Society, the so-called Stoddard Swamp has become the precious state-owned Lucy Mallary Bugbee Natural Area.

In the Northeast, peatlands usually comprise minor topics within broader natural areas programs. Systematic efforts on state or regional levels now in their infancy may grow to provide significant, perhaps even coordinated, peatland preservation throughout this nine-state region. At the state level Maine has led the way; its Critical Areas Program has specifically addressed peatlands for the past decade, producing a bevy of researches, reports, assessments, and registered sites. In other states The Nature Conservancy, through its Heritage Program and with assistance of some state agencies, is developing a systematic approach to identifying the peatlands most appropriate for preservation because of their contribution to natural diversity. Federal involvement in the deliberate protection of naturally significant peatlands remains minimal. The poorly funded but competently staffed National Natural Landmark Program of the National Park Service has registered as a National Natural Landmark at least one peatland in each northeastern state; Maine alone has six, Pennsylvania five.

In addition to the systematic programs, protection of peatlands comes from various sources, all almost totally independent of each other. These include private owners, local citizens' and sportsmen's groups, amateur naturalist groups, statewide and national conservation organizations, quasipublic institutions (such as universities, colleges, foundations, and religious institutions), and public agencies at all levels. Some of this protection stems from the peatlandness (usually the bogness) of the site; some comes incidentally as a larger site is preserved for other reasons—for example, the University of Vermont Natural Areas contains two bog sites plus an alpine summit that includes many acres of tundra as well as two tiny peatlands. Several U.S. Fish and Wildlife Refuges in the Northeast, in their management of fish and bird habitats, maintain peatlands in their natural state.

So at least we have begun. We have taken notice; we are studying; we are peering into the future. We discover untouched peatlands; we see peatlands lost. We see the need for balance between utilization and preservation. It is the modern dilemma all over again in another arena—commercial exploitation versus irretrievable loss. Where does use outweigh loss? Where does use leave only gentle footsteps? Where must use not be? Where must the sacrifices be?

We must never lose sight of the fact that we have only a limited number of peatlands. We cannot make them, nor wait around another 5,000 or 10,000 years for other peatlands to replace ones trucked away and burned. We must take care of what we already have: save the best, the most significant, the rare, the local—care for and respect them in this day and in centuries to come. And for all natural peatlands this thought concluding the prologue to *Maine Peatlands* (Worley 1981) rings true:

Living landscapes, once quiet, seclusive and seldom visited, peatlands in Maine can now expect the full force of burgeoning society in search of untapped resources to force their exposé. Indigenous ecosystems, here nearly untouched by man, soon to have their usefulness and very fate determined by the decisions of the single species *Homo sapiens*, peatland deserves the rational thought of man whose teleology is as capable of mortal error as it is of joyous achievement.

Though we have come a long way in gaining knowledge of peatlands, we still do not know enough to answer all the important questions that have been asked. Maybe we don't even know all the questions that should be asked. But at least we do know that peatlands are not just strange blotches on the landscape, nor useless, nor unimportant. Rather, they store treasures of our natural heritage and provide refuge for rare species of our earth. Where abundant, they can give us fuels and products we seek in this demanding age. They can be magical inspirations for the minds and souls of people. And they, like tall mountains above, broad seas beyond, or nestling forests that surround, should claim some autonomous recognition, apart from our personal or selfish judgments—for simply being here, with us, on earth together.

The pleas are eloquent. A recent major article in *Audubon* about proposed mining bears this gripping title by author John Luoma: "The Big Bog: Requiem for a Lonely Wilderness."

In the epilogue of his recent book on North American peat-

lands James Larsen prematurely dedicates an epitaph, in hopes
no final passing shall ever come to be:

In this book I have attempted to capture the essence of a distinctive
native plant community in what may be the last fleeting moments
before final extinction. It seems almost inevitable that in the northern United States the ombrotrophic bogs will be crushed beneath
the advance of human populations unless forceful action is taken
soon to preserve them. . . . This, then, is intended to be a record of
the last of the bogs.

Ian Worley, before a Maine conference on the preservation
of peatlands, appealed:

Let the rich variety of this wee globe live, and thrive in mosaic with
species man—let our aged, grey-bearded ecosystems preserve their
waterlogged cedar chests of antiquities for another 10,000 years!

Reasons there are enough for us to watch over the Northeast's peatlands, from Cape May to the Allagash, from Lake
Erie to the Gulf of Maine—to cherish them as gifts and fel-
low travelers on earth's odyssey—to allow some to exist on
their own, to go where they will. We owe as much to this
planet and to all its children yet to come.

BIBLIOGRAPHY

The following references are by no means a complete list of important works dealing with peatlands. Rather, they serve as a broad introduction to the subjects discussed in this book. For specialists and those who want to investigate one subject in more detail, the bibliographies in Gore (1983), Worley (1981), Radforth and Brawner (1977), and Moore and Bellamy (1974) are good starting places. To obtain hard-to-find and newly published materials, contact the reference librarian of any college or university library.

Andrus, R. E. 1980. *Sphagnaceae (peat moss family) of New York State.* New York State Museum Bulletin 442, Albany.

Behler, J. L., and F. W. King. 1979. *The Audubon Society field guide of North American reptiles and amphibians.* New York: Alfred A. Knopf.

Billings, W. D., and N. A. Mooney. 1968. The ecology of arctic and alpine plants. *Biological Review* 43:481–529.

Bishop, S. C. 1947. *Handbook of salamanders.* Ithaca, N.Y.: Comstock.

Bliss, L. C. 1963. Alpine plant communities of the Presidential Range, New Hampshire. *Ecology* 44:678–97.

Boelter, D. H., and E. S. Verry. 1977. *Peatland and water in the northern lake states.* U.S. Forest Service General Technical Report NC-31. St. Paul, Minn.: North Central Forest Experimental Station.

Borror, D. J., and R. E. White. 1970. *A field guide to the insects of America north of Mexico.* Boston: Houghton Mifflin.

Bradstreet, T. E., and R. B. Davis. 1975. Mid-postglacial environments in New England with emphasis on Maine. *Arctic Anthropology* 12:7–22.

Brande, J. 1979. Worthless, valuable, or what? An appraisal of wetlands. *Journal of Soil and Water Conservation* 35:12–16.

Brewer, R. 1967. Bird populations of bogs. *Wilson Bulletin* 79:371–96.

Brower, A. E. 1978. *Bog elfin,* Incisalia lanoiaieensis, *in Maine and its relevance to the Critical Areas Program.* Critical Areas Program, Report no. 63. Augusta: Maine State Planning Office.

Burke, I. 1982. *The Mahoosuc Mountains: A natural areas inventory and management statement.* Report prepared for the Maine Department of Conservation by the Maine Critical Areas Program, State Planning Office and the Appalachian Mountain Club. Augusta and Boston.

Burt, W. H., and R. P. Grossenheider. 1964. *A field guide to the mammals.* Boston: Houghton Mifflin.

Caljouw, C. A. 1982. *The Great Heath: A natural areas description.* Augusta: Executive Department, Bureau of Public Lands and Maine State Planning Office.

Conant, R. 1975. *A field guide to reptiles and amphibians of eastern and central North America.* Boston: Houghton Mifflin.

Cornell, D. S. 1950. *Native orchids of North America north of New Mexico.* Stanford, Calif.: Stanford University Press.

Cowardin, L. M., V. Carter, F. C. Golet, and E. T. LaRoe. 1979. *Classification of wetlands and deep-water habitats of the United States.* Washington, D.C.: U.S. Department of the Interior, Fish and Wildlife Service.

Cox, C. B., I. N. Healey, and P. D. Moore. 1976. *Biogeography.* Oxford: Blackwell Scientific Publications.

Crow, G. E., W. D. Countryman, G. L. Church, L. M. Eastman, C. B. Hellquist, L. L. Mehrhoff, and I. M. Storks. 1981. Rare and endangered vascular plant species in New England. *Rhodora* 83(834):259–99.

Crum, H. A., and L. E. Anderson. 1981. *Mosses of eastern North America.* 2 vols. New York: Columbia University Press.

Dachnowski, A. P. 1926. Profiles of peat deposits in New England. *Ecology* 7:120–35.

Damman, A. W. H. 1977. Geographical changes in the vegetation pattern of raised bogs in the Bay of Fundy region of Maine and New Brunswick. *Vegetatio* 35(3):137–51.

———. 1978. Distribution and movement of elements in ombrotrophic peat bogs. *Oikos* 30:480–95.

———. 1979. Geographic patterns in peatland development in eastern North America. In *Proceedings of the International Symposium on Classification of Peat and Peatlands.* Hyytiala, Finland: International Peat Society.

Dansereau, P., and F. Segadas-Vianna. 1952. Ecological study of the peat bogs of eastern North America. I: Structure and evolution of vegetation. *Canadian Journal of Botany* 30:490–520.

Daubenmire, R. 1978. *Plant geography with special reference to North America.* New York: Academic Press.

Davis, J., and G. K. White. 1979. *Production and utilization of Maine's peat resources.* Orono: University of Maine, Department of Agriculture and Resource Economics.

Davis, R. B., G. L. Jacobson, L. S. Widoff, and A. Zlotsky. 1983. *Evaluation of Maine peatlands for their unique and exemplary qualities. A report to the Maine Department of Conservation (draft).* Orono: University of Maine, Institute for Quaternary Studies and Department of Botany and Plant Pathology.

Deevey, E. S., Jr. 1958. Bogs. *Scientific American* 199:115–21.
DeGraaf, R. M., and D. D. Rudis. 1983. *Amphibians and reptiles of New England*. Amherst: University of Massachusetts Press.
Densmore, F. 1974. *How Indians used wild plants for food, medicine, and crafts*. Reprint. New York: Dover.
Dowden, A. O. 1975. To pollinate an orchid. *Audubon* 77(5):40–47.
Dressler, R. L. 1981. *The orchids: Natural history and classification*. Cambridge, Mass.: Harvard University Press.
Erdman, K. S., and P. G. Wiegman. 1974. *Preliminary list of natural areas in Pennsylvania*. Pittsburgh: Western Pennsylvania Conservancy.
Ernst, C. H., and R. W. Barbour. 1972. *Turtles of the United States*. Lexington: University Press of Kentucky.
Farnham, R. S. 1978. Wetlands as energy sources. In *Wetland functions and values: The state of our understanding. Proceedings of the National Symposium on Wetlands*. Minneapolis: American Water Resources Association.
Fenneman, N. M. 1938. *Physiography of eastern United States*. New York: McGraw-Hill.
Gemming, E. 1983. *The cranberry book*. New York: Coward-McCann.
Gleason, H. A., and A. Cronquist. 1963. *Manual of vascular plants of northeastern United States and adjacent Canada*. Princeton, N.J.: D. Van Nostrand.
———. 1964. *The natural geography of plants*. New York: Columbia University Press.
Glob, P. V. 1969. *The bog people: iron age man preserved*. Ithaca, N.Y.: Cornell University Press.
Godin, A. J. 1977. *Wild mammals of New England*. Baltimore: Johns Hopkins University Press.
Godwin, Sir H. 1981. *The archives of the peat bogs*. Cambridge: Cambridge University Press.
Good, R. E., D. F. Whigham, and R. L. Simpson. 1978. *Freshwater wetlands: Ecological processes and management potential*. New York: Academic Press.
Goodwillie, R. 1979. *European peatlands*. Strasburg: European Committee for the Conservation of Native and Natural Resources, Council of Europe. Mimeographed report.
Goodwin, R. H., and W. A. Niering. 1975. *Inland wetlands of the United States*. Natural History Theme Studies, no. 2. Washington, D.C.: National Park Service.
Gore, A. J. P., ed. 1983. *Ecosystems of the world. Mires: Swamp, bog, fen, and moor*. Parts 4A and 4B. Amsterdam: Elsevier.
Gorham, E. 1957. The development of peat land. *Quarterly Review of Biology* 32:145–66.
———. 1983. The ecology and biochemistry of *Sphagnum* bogs. Third Annual Report, Minnesota Peat Project, St. Paul.
Harshberger, J. W. 1970. *The vegetation of the New Jersey Pine Barrens*. New York: Dover.

Heinselman, M. L. 1970. Landscape evolution, peatland types, and the environment in the Lake Agassiz peatlands natural area, Minnesota. *Ecological Monographs* 40:235–61.

Hemond, H. F. 1980. Biogeochemistry of Thoreau's Bog, Concord, Massachusetts. *Ecological Monographs* 50:507–26.

Henry, L. K. 1950. Comparison of the floras of some western Pennsylvania bogs. *Proceedings of the Pennsylvania Academy of Science* 24:21–25.

———. 1963. Some interesting bogs in Erie County, Pennsylvania. *Trillia* 12:22–24.

Hodge, W. H. 1981. Where a heavy body is likely to sink. *Audubon* 83(5):98–111.

Hyland, F., and B. Hoisington. 1977. *The woody plants of sphagnous bogs of northern New England and adjacent Canada.* Maine Bulletin 744. Orono: University of Maine, Life Science and Agricultural Experiment Station.

Ingram, H. A. P. 1967. Problems of hydrology and plant distribution in mires. *Journal of Ecology* 55(3):711–24.

Jacobson, G. L., Jr., and L. Widoff. 1983. The vegetation of Maine peatlands, with special reference to Caribou Bog, Crystal Bog, and the Great Heath. Report to the Maine Geological Survey, Augusta.

Johnson, C. W. 1980. *The nature of Vermont: Introduction and guide to a New England environment.* Hanover, N.H.: University Press of New England.

Jones, F. M. 1921. Pitcher plants and their moths. *Natural History* 21(3): 296–316.

Jorgensen, N. 1977. *A guide to New England's landscape.* Chester, Conn.: Pequot Press.

———. 1978. *A Sierra Club naturalist's guide to southern New England.* San Francisco: Sierra Club Books.

Karns, D. R. 1979. *The relationship of amphibians and reptiles to peatland habitats in Minnesota.* St. Paul: Minnesota Peat Program, Division of Minerals, Department of Natural Resources.

———. 1981. *Bog water toxicity and amphibian reproduction.* St. Paul: Minnesota Peat Program, Division of Minerals, Department of Natural Resources.

Kivinen, E., and Pakarinen, P. 1981. Peatland areas and the proportion of virgin peatlands in different countries. In *Proceedings of the Sixth International Peat Congress,* Duluth, Minnesota.

Klots, A. B. 1951. *A field guide to butterflies of North America east of the Great Plains.* Boston: Houghton Mifflin.

Kovacs, R. F. 1979. The Algonkian Indians' use of some of the plants indicative of a bog habitat. Class paper, Botany Department, University of Vermont, Burlington.

Kulik, S., P. Salmansohn, M. Schmidt, and H. Weich. 1984. *The Audubon Society field guide to the natural places of the Northeast: Coastal and inland.* 2 vols. New York: Pantheon.

Larsen, J. A. 1982. *Ecology of the northern lowland bogs and conifer forests.* New York: Academic Press.

Lawrence, S., and B. Gross. 1984. *The Audubon Society field guide to the natural places of the mid-Atlantic states: Coastal and inland.* 2 vols. New York: Pantheon.

Lazell, J. D., Jr. 1976. *This broken archipelago.* New York: Demeter Press, Quadrangle, New York Times Book Co.

Lloyd, F. E. 1942. *The carnivorous plants.* Reprint. New York: Dover, 1976.

Luer, C. A. 1975. *The native orchids of the United States and Canada (excluding Florida).* New York: New York Botanical Garden.

Lull, H. W. 1968. *A forest atlas of the Northeast.* Upper Darby, Pa.: Northeastern Forest Experimental Station, U.S. Forest Service.

Luoma, J. 1981. The big bog: Requiem for a lonely wilderness. *Audubon* 83(5):112–27.

Lyon, C. J., and W. A. Reiners. 1971. *Natural areas of New Hampshire.* Dartmouth College, Department of Biological Sciences, Publication no. 4. Hanover, N.H.

Maine Department of Inland Fisheries and Wildlife. 1982. *Fisheries and potential impacts of peat mining in Maine.* Augusta.

Marchand, P. J. 1977. *Subalpine bogs of the Mahoosuc Range, Maine: Physical characteristics and vegetation development.* Wolcott, Vt.: Center for Northern Studies.

Marshall, W. H., and D. G. Miquelle. 1978. *Terrestrial wildlife of Minnesota peatlands.* St. Paul: Minnesota Peat Program, Minnesota Department of Natural Resources.

May, D. E., and R. B. Davis. 1978. *Alpine tundra vegetation on Maine mountains and its relevance to the Critical Areas Program.* Maine Critical Areas Program, Report no. 36. Augusta: Maine State Planning Office.

McElroy, T. P., Jr. 1974. *The habitat guide to birding.* New York: Alfred A. Knopf.

McVaugh, R. 1957. *Flora of the Columbia County Area, New York.* New York State Museum and Science Service Bulletin 360. Albany.

Minnesota Department of Natural Resources, Division of Minerals. 1981. *Minnesota peat program final report.* St. Paul.

Moore, P. D., and D. J. Bellamy. 1974. *Peatlands.* New York: Springer-Verlag.

The Nature Conservancy. 1977. *Preserving our national heritage.* 2 vols. Arlington, Va.

New Jersey Pinelands Commission. 1980. *Comprehensive management plan for the Pinelands National Reserve.* New Lisbon, N.J.

New York Energy Research and Development Authority. 1982. *State of New York peat resource inventory.* Albany.

Niering, W. A. 1953. *The past and present vegetation of High Point State Park. Ecological Monographs* 23:127–48.

————. 1966. *The life of the marsh: The North American wetlands.* New York: McGraw-Hill.

Pease, A. S. 1964. *A flora of northern New England.* Cambridge, Mass.: New England Botanical Club.

Perry, J., and J. G. Perry. 1980. *The Random House guide to natural areas of the eastern United States.* New York: Random House.

Peterson, B. S., C. E. Cross, and N. Tilder. 1968. *The cranberry industry in Massachusetts.* Massachusetts Department of Agriculture Bulletin 201, Boston.

Peterson, R. T. 1980. *A field guide to the birds.* Boston: Houghton Mifflin.

Pettingill, O. S., Jr. 1977. *A guide to bird finding east of the Mississippi.* New York: Oxford University Press.

Pietz, P. J., and J. R. Tester. 1979. *Utilization of Minnesota peatland habitats by snowshoe hare, white-tailed deer, spruce grouse, and ruffed grouse.* St. Paul: Minnesota Peat Program, Division of Minerals, Department of Natural Resources.

Pollett, F. C., A. F. Rayment, and A. Robertson, eds. 1980. *The diversity of peat.* St. John's: Newfoundland and Labrador Peat Association.

Radforth, N. W., and C. O. Brawner, eds. 1977. *Muskeg and the northern environment in Canada.* Toronto: University of Toronto Press.

Robichaud, B., and M. F. Buell. 1973. *Vegetation of New Jersey: A study of landscape diversity.* New Brunswick, N.J.: Rutgers University Press.

Schnell, D. E. 1976. *Carnivorous plants of the United States and Canada.* Winston-Salem, N.C.: John F. Blair.

Seymour, F. C. 1969. *The flora of New England.* Rutland, Vt.: Charles E. Tuttle.

Shelter, S. G., and F. Montgomery. 1965. *Insectivorous plants.* Smithsonian Institution Leaflet no. 447, Washington, D.C.

Sheridan, E. T. 1965. *Peat. Minerals yearbook.* Washington, D.C.: U.S. Department of the Interior.

Siccama, T. G., W. A. Beales, and J. E. Hibbard. 1973. *Connecticut natural areas.* East Hartford: Connecticut Forest and Park Association.

Slack, A. 1980. *Carnivorous plants.* Cambridge, Mass.: MIT Press.

Stockwell, S. S., and M. L. Hunter. 1983. A description of the wildlife in twenty-seven Maine peatlands. Report to the Maine Department of Inland Fisheries and Wildlife, Augusta.

Thomson, B. F. 1977. *The changing face of New England.* Boston: Houghton Mifflin.

Tyler, H. R., Jr., and C. V. Davis. 1982. *Evaluation of No. 5 Bog and jack pine stand, Somerset County, Maine, as a potential national natural landmark.* Augusta: Critical Areas Program, Maine State Planning Office.

U.S. Department of Energy, Division of Fossil Fuels Processing. 1979. *Peat prospectus.* Washington, D.C.

Van Der Pijl, L., and C. H. Dodson. 1966. *Orchid flowers: Their pollination and evolution.* Coral Gables, Fla: University of Miami Press.

Vogelmann, H. W. [1964] 1971. *Natural areas in Vermont.* Report 1, Vermont Agricultural Experiment Station, Burlington, Vt.

————. 1969. *Vermont natural areas.* Report 2, Central Planning Office and Interagency Committee on Natural Resources, Burlington, Vt.

Weiner, M. A. 1980. *Earth medicine—earth food: Plant remedies, drugs, and natural foods of the North American Indians.* Rev. ed. New York: Macmillan, Collier.

Whittaker, R. H. 1970. *Communities and ecosystems.* New York: Macmillan.

Worley, I. A. 1980. *Botanical and ecological aspects of coastal raised peatlands in Maine and their relevance to the Critical Areas Program of the State Planning Office.* Critical Areas Program, Report no. 69. Augusta: Maine State Planning Office.

————. 1981. *Maine Peatlands: Their abundance, ecology, and relevance to the Critical Areas Program.* Augusta and Burlington: Critical Areas Program, Report no. 73, Maine State Planning Office; and Bulletin 687, Vermont Agricultural Experiment Station.

Worley, I. A., and R. Klein. 1980. Protection and preservation of peatlands as natural areas in the northeastern United States. In *Proceedings of the Sixth International Peat Congress,* Duluth, Minn.

Worley, I. A., and J. R. Sullivan. 1980. *A classification scheme for the peatlands of Maine.* Vermont Agricultural Experiment Station, Resource Report 8, Burlington, Vt.

Zahl, P. A. 1961. Plants that eat animals. *National Geographic* 119: 642–59.

APPENDIX 1 | SOME PEATLANDS OF THE NORTHEAST

This appendix offers a closer look at some significant or representative peatlands of the region. Many of these I have explored personally; others I have not visited but have included here because of their importance. All are representative of major peatland types in each state, or are representative of different geographical regions of each state, or have recognized natural significance (usually at the state or federal level).

Certain peatlands with considerable natural significance are not included here in order to protect their vulnerable, fragile, easily destroyed features. Owing to the rarity and sensitivity of rich fens in the Northeast and to the endangered status of many of their plants, none has been identified in the following accounts unless it is large and remote enough that its crucial parts are adequately protected.

Visitors should remember that *all* peatlands are in a sense fragile areas and should be treated with utmost care during exploration. The following are "canons" of bog etiquette.

• Contact the owner and follow regulations and/or requests.
• Do not pick or uproot flowers or plants unless you have express permission to do so. Even then, be judicious.
• Step lightly and with care.
• If you go in a group, keep it small.
• Of course, do not litter.

Many peatland features are most easily seen during winter months when frozen ground permits easier traveling. Likewise, winter visits cause much less damage than visits during other seasons.

Those who have never visited a peatland, or only at one season, should also heed the following words of advice. Though peatlands are not really treacherous, as they often are rumored to be, they can present some difficulties to those who are careless or unused to them.

- Laggs (when they occur) are often the wettest sections of many peatlands, so don't get discouraged if you find entry somewhat difficult. The bog usually gets drier farther in.
- Be prepared for dense, tall shrub thickets, especially at or near margins of peatlands.
- Fens tend to be much more difficult to explore than bogs because they are wet throughout.
- As you walk through a peatland, keep to the areas where the mat is firm. Avoid or be careful on the very unsteady pond edges and dark, unvegetated peaty areas. In both places you can literally get bogged down!
- In summer: Be prepared for biting and stinging insects—black flies, mosquitoes, and/or deer flies, depending on the month and area.
- Wear old sneakers; rubber boots may get too hot. Open bogs especially can be hot in summer.
- In early spring or late fall: The waters may be cold, so wear rubber boots (knee or hip length) and warm clothing. Chest waders are too cumbersome.
- In *any* season, but especially in the cold months, have a change of clothing back in the car. Footwear and socks are most important.
- Depending on your interests, take along guidebooks, binoculars, camera, and perhaps an enthusiastic naturalist—all helping you better to see, understand, appreciate, and respect these fascinating but fragile ecosystems.

Although most of the peatlands mentioned here are public properties, parts may be in private ownership, have access through private grounds, or in some way involve a landowner. *Please* respect the rights and privacy of that landowner by honoring signs and following normal courtesies. If you have doubts about whether you are allowed in certain places, the best policy is either not to go there or to ask permission. Most publicly owned peatlands have protective regulations and guidelines. Do contact the responsible agencies and follow the proper practices. When no ownership is indicated for sites in this list, there is usually more than one private owner.

In the following, acreages are given for most sites. In many cases these are approximations at best but are included to give the general size of the peatland, or in some cases, the management area containing the peatland.

Following the entries for each state is a list of agencies to contact for additional information. The organizations below are good sources of information about the region as a whole.

Notes: States are listed roughly in a south-to-north, west-to-east progression; bogs are listed alphabetically in each state. Scientific names of plants and animals are not given; all species are mentioned elsewhere in the book (see index).

REGIONAL CONTACTS AND FURTHER INFORMATION

National Park Service
Mid-Atlantic Region
143 South Third Street
Philadelphia, Pennsylvania 19106

National Park Service
North Atlantic Region
15 State Street
Boston, Massachusetts 02109

The Nature Conservancy
Eastern Regional Office
294 Washington Street
Boston, Massachusetts 02108

U.S. Fish and Wildlife Service
Region 5
One Gateway Center
Newton Corner, Massachusetts 02158

U.S. Forest Service
Eastern Region
310 West Wisconsin Avenue
Milwaukee, Wisconsin 53203

NEW JERSEY

Great Swamp, Great Swamp National Wildlife Refuge, near New Providence. A National Natural Landmark. Owned by the U.S. Fish and Wildlife Service. Within this 5,500-acre complex of wetlands are several peatlands, all part of an old glacial lake, Lake Passaic. The peatlands, too, are varied. Several have been disrupted by drainage for agriculture and are generally oligotrophic to weakly minerotrophic; sedges, red maples, leatherleaf, rhododendron, and cattails predominate. In acidic sections of the swamp, heath shrubs become more abundant. Many species of birds, amphibians, and reptiles are present throughout. Access is from several locations around and through the swamp; a nature center serves as a good starting and orientation point, providing a boardwalk partway into the wetlands.

Kuser Memorial Natural Area (Cedar Bog), High Point State Park, near Sussex. Owned by the New Jersey Department of Environmental Protection. The 200-acre Atlantic white cedar swamp and peatland that forms part of this large state park is protected as a natural area. A marginal trail and an old woods road bisecting the peatland offer good access. Mostly for-

Fig. 69 A Pine Barrens leatherleaf "spong." *(Photo by Robert A. Zampella.)*

ested with Atlantic white cedar, black spruce, hemlock, along with some black gum, white pine, red maple, and yellow birch, this poor fen has a rich understory of rhododendron, skunk cabbage, ferns, and other herbs. Canopy breaks foster sundew and pitcher plant growth. Rich in animal life, there can be seen many bird and mammal species.

Lily Lake Natural Area, Brigantine National Wildlife Refuge, near Oceanville. Owned by the U.S. Fish and Wildlife Service. This refuge, famous for its variety of wildlife and wetlands, includes a 3-acre bog characterized by Atlantic white cedar and red maple, with an understory of heaths and other shrubs and a ground cover of sphagnum mosses. The site is important as a home of the Pine Barrens tree frog.

Pine Barrens. The vast public and private million acres of the New Jersey Pine Barrens contain many peatlands, whose characteristics have been described in Chapter 5. Generally, the bogs are quite acidic and form, in conjunction with streams or springs in old pond beds, on sandy soils. The peats are shallow, overtopped by mats of sphagnum, sedges, sundews, pitcher plants, and masses of leatherleaf. The bogs, often formerly cut-over white cedar swamps and/or abandoned cranberry bogs, are noteworthy for their variety of plants, many with predominantly southern distributions (see page 33). Some of the best examples of these special places are in Wharton State Forest, Penn State Forest, Bass River State Forest, and Lebanon State Forest, all owned by the New Jersey Department of Environmental Protection. Many of the peatlands in these state forests have been given natural area status and protection. For several of these publicly owned lands maps and checklists of plants and animals are available. Wharton State Forest also has a nature center.

Pine Bog (Pine Swamp), High Point State Park. Owned by the New Jersey Department of Environmental Protection. One of two bogs in the state park, Pine Bog is the deepest, with seventeen feet of peat accumulation. The floating mat surrounding Lost Lake has a rather typical assemblage of plants: sphagnum mosses, heaths (bog rosemary, bog laurel, small cranberry), and some stunted black spruce and tamarack. More mineral-rich sections have red maple, black gum, highbush blueberry, rhododendron, speckled alder, swamp azalea, and others.

Wawayanda Cedar Swamp, Wawayanda State Park, near Vernon. Owned by the New Jersey Department of Environmental Protection, Division of Parks and Forestry. This swamp-peatland in the extreme northern part of the state is designated as a natural area and comprises more than 2,000 acres. It has Atlantic white cedar and other plants typical of this kind of landscape. The peatland sections are small.

Webb's Mills Bog, Lacey Township. Owned by the New Jersey Department of Environmental Protection, Division of Fish, Game, and Wildlife. This 6- to 10-acre Atlantic white cedar peatland also has a quaking mat of sphagnum, sedges, sundews, some shrubs, and pitcher plants, along with some open water areas.

Contacts and Further Information

New Jersey Audubon Society
1790 Ewing Avenue
Franklin Lakes, New Jersey
07417

New Jersey Conservation
Foundation
300 Mendham Road
Morristown, New Jersey 07960

New Jersey Pinelands
Commission
P.O. Box 7
New Lisbon, New Jersey 08064

New Jersey Department of
Environmental Protection
Division of Parks and Forestry
P.O. Box 1420
Trenton, New Jersey 08625

(This department includes the Office of Natural Lands Management, which has incorporated the former Green Acres Program, part of the New Jersey Natural Heritage Program.)

Fig. 70 Bear Meadows, Pennsylvania. *(Photo by Charles W. Johnson.)*

PENNSYLVANIA

Bear Lake, Warren County, near Bear Lake. 15 acres. Kettlehole peat-lands such as this are uncommon in Pennsylvania because so few areas in the state were glaciated. The almost circular pond here is ringed with typi-cal northern bog shrubs such as leatherleaf, winterberry, mountain holly, and chokeberry. Back from the water's edge, the bog becomes more minero-trophic, probably influenced by water inlets; plants such as skunk cabbage and Virginia chain fern are common in this section.

Bear Meadows Natural Area, near State College. Owned by the Pennsyl-vania Department of Environmental Resources. The entire natural area, a National Natural Landmark, is 833 acres, more than half of which is peat-land. This poor fen is important also for the wildlife in and around it, including beaver, white-tailed deer, and plentiful songbirds. The upland-peatland border is a minor jungle of rhododendron, giving way to highbush blueberry farther out on the peat soils. The main, open part of the peatland is dominated by robust growth of leatherleaf and highbush blueberry, the former often reaching chest height. Sluggish streams move through this fen, and beaver are active in several places. Access is by good trails, and an ob-servation platform in the peatland gives fine views while minimizing visitor disturbance.

Bear Swamp, Tobyhanna State Park, near Tobyhanna. Owned by the Pennsylvania Department of Environmental Resources. The peatland appears to cover about 100 acres in the northeastern quarter of the state. Shrubby areas thick with highbush blueberry and rhodora, along with mountain holly, black chokeberry, and several other species, dominate the peatland. Open "trails" thread through the thickets; here grow sphagna, white beak-rush, sundews, cotton grasses, and stunted gray birch.

Crystal Springs Bog, S. B. Elliott State Park, near Goshen. Partly owned by the Pennsylvania Department of Environmental Resources. This peatland of about 30 acres provides a good example of how a bog can change following severe disturbance. It was logged off earlier in the century, then mined for peat. Since abandonment bog plants have been reestablishing. Cattails, growing where the peat was removed and water took its place, now are being squeezed out by burgeoning sphagnum mosses. White beak-rush, round-leaved sundews, and black chokeberry are particularly prevalent amid larger sphagnum patches. Bog clubmoss is locally abundant, and some black gum trees grow at the margins of the bog. The peatland can be observed easily from the old haul road through it, built for the earlier mining operations.

Hartstown Bog, Pymatuning State Gameland, near Hartstown. Owned by the Pennsylvania Department of Environmental Resources. Unfortunately, the original peatlands and their associated ponds, spanning more than 3,000 acres, have been inundated with water levels raised to promote waterfowl production on the state wetlands, of which the peatlands were a part. However, several large sections of both wooded and open peatland still exist and are worth exploring, as are the other wetlands of this huge reserve.

Pine Lake Natural Area, near Greentown. Owned by the Pennsylvania Department of Environmental Resources. This 67-acre peatland, situated on the tabletop plateau of the Pocono Mountains, displays zonation from open water to a sweet gale-leatherleaf margin to a sphagnum-sedge main mat with a tamarack fringe inland.

Reynolds' Spring Natural Area, near Cedar Run. Owned by the Pennsylvania Department of Environmental Resources. The state forest natural area here consists of 1,302 acres, with a peatland occupying perhaps 100 acres. At a relatively high-elevation site in the midst of a white pine-pitch pine-oak forest, this somewhat minerotrophic peatland's sphagnum mat harbors numerous intriguing species, including round-leaved sundew, pitcher plant, buckbean, and a variety of colorful orchids.

Rosecrans Bog Natural Area, near Rosecrans. Owned by the Pennsylvania Department of Environmental Resources. Another peatland in the mountains of central Pennsylvania, this 152-acre natural area has plant species typical of the region, including tall shrubs of mountain holly and highbush bluebery, with understories and openings of sphagnum mosses, sedges, cranberries, and a variety of additional poor fen species of herbs and dwarf shrubs.

Tannersville Cranberry Bog Preserve, near Tannersville. *Special note*: Permission is necessary for entry to this privately owned preserve. Contact the Meesing Nature Center, c/o Monroe Conservation District, RD 2, Box 2335-A, Stroudsburg, Pennsylvania 18360, (717) 992-7334. This fine area is somewhat difficult to get through but highly rewarding. A moist lagg, lush with rhododendron, marsh and cinnamon ferns, burreed, poison sumac, and maleberry, borders the approximately 300-acre poor fen. In the mostly wooded peatland grow black spruce, red maple, yellow birch, rhododendron, and poison sumac over a relatively rich and varied understory.

Wattsburg Fen (Weber's Bog), near Wattsburg. Owned by the Western Pennsylvania Conservancy. Permission required to enter. This 25-acre fen, arising from the calcareous bedrock of lakeside Erie County, has species typical of both poor and richer fens, including several orchids. Although not large, Wattsburg Fen supports a mixture of sphagnum, heath shrub, and open water communities.

Contacts and Further Information

The Nature Conservancy
Pennsylvania/New Jersey Office
1218 Chestnut Street, Suite 1002
Philadelphia, Pennsylvania 19107

Western Pennsylvania
Conservancy
316 Fourth Avenue
Pittsburgh, Pennsylvania 15222

Pennsylvania Department of
Environmental Resources
Press Office
Fulton Building, 9th Floor
Harrisburg, Pennsylvania 17120

CONNECTICUT

Beckley Bog (Walcott Preserve), near Norfolk. Owned by The Nature Conservancy, Connecticut Chapter. "One of the finest northern-type bogs in the state" (Goodwin and Niering, 1975). Its 600 acres include a 7-acre lake; a forested poor fen with such species as black spruce, tamarack, heaths, carnivorous plants; and an open sedge fen. Located in the low mountains of the Northwestern Highlands, the whole peatland/wetland complex is confined to a narrow valley.

Bingham Pond Bog, near Salisbury. This 60-acre pond in the highlands of northwestern Connecticut has a peatland at one end containing shrubs (predominantly leatherleaf), tamarack, and black spruce. The bog is one of several wetlands associated with the valleys of this hilly terrain.

Fig. 71 Black Spruce Bog, Connecticut.
(Photo by Charles W. Johnson.)

Black Spruce Bog, Mohawk Forest, near Goshen. Owned by the Connecticut Department of Environmental Protection. This 10- to 15-acre black spruce peatland occupies a glacially carved depression atop a modest mountain. A good boardwalk leads through a hemlock-mountain laurel border into the peatland proper, where black spruce and tamarack predominate. The dense understory has much highbush blueberry and mountain holly. Sphagnum is a ground cover throughout.

Cranberry Pond Bog, near Litchfield. Owned by the White Memorial Foundation. This approximately 50-acre beaver-influenced wetland and lake has portions of peatland consisting of black spruce, tamarack, pitcher plants, huckleberry, winterberry, highbush blueberry, and others. Other portions are more marshy, lush with sedges and cattails. Studies here have shown the peat deposits to be as much as 48 feet deep.

Haley Farm Bog, Haley Farm State Park, near Groton. Owned by the Connecticut Department of Environmental Protection. This 2-acre pond peatland is an interesting example of change. Once a tidal marsh, the area was cut off from saltwater access by a private racetrack and railroad line, and a peatland began developing where fresh water was impounded. Freshwater peats have been deposited on top of former saltwater marsh peats. Now the surface supports a poor fen flora of sundews, sedges, buckbean, sphagnum mosses, and orchids and is bordered by speckled alder. Trails provide easy access.

Little Pond, near Litchfield. Owned by the White Memorial Foundation. A long elevated boardwalk leads through and above this varied and interesting 50-acre wetland. Before European colonization an Atlantic white cedar, black spruce, and tamarack peatland, fluctuations in water level have resulted in a shrub swamp, with clear zonation from open water to buttonbush to willows, swamp roses, and cattails.

Pachaug Great Meadow, near Voluntown (600 acres), and *Cedar Swamp*, near Chester (400 acres), are large and noteworthy Atlantic white cedar swamps/peatlands, with typical plant communities described in Chapter 5.

Contacts and Further Information

Connecticut Audubon Society
613 Riverville Road
Greenwich, Connecticut 06833

Connecticut Department of
Environmental Protection
State Office Building
165 Capitol Avenue
Hartford, Connecticut 06115

Connecticut Geological and
Natural History Survey
c/o Biological Services Group
University of Connecticut,
Box 4–42
Storrs, Connecticut 06268

Connecticut Natural Heritage
Program
Natural Resources Center
Department of Environmental
Protection
State Office Building, Room 561
Hartford, Connecticut 06115

The Nature Conservancy,
Connecticut Chapter
Box MMM Wesleyan Station
Middletown, Connecticut 06457

White Memorial Foundation, Inc.
Box 368
Litchfield, Connecticut 06759

RHODE ISLAND

Arcadia Black Spruce Bog, Arcadia Management Area, near West Exeter. Owned by the Rhode Island Department of Environmental Management. Black spruce, which is at its southern extremity in the state here, shares the 5-acre bog with dwarf shrubs and other bog plants. Access by trail.

Diamond Bog, Carolina Management Area, near Woodville. Owned by the Rhode Island Department of Environmental Management. This wetland complex is fairly large as Rhode Island bogs go, comprising more than 60 acres. There are several peatland types, including a quaking bog of sphagnum mosses, shrubs, and herbaceous plants; an Atlantic white cedar

Fig. 72 Diamond Bog, Rhode Island. *(Photo by Rick Enser.)*

swamp; and a sedge fen. The flora and fauna have been extensively studied by scientists at the University of Rhode Island.

Ell Pond, near Hope Valley. Several peatlands occur in this pond-rich section of Rhode Island. All are of rather typical quaking-bog composition with heath-sedge-sphagnum mat and associated herbs. It is perhaps 50 acres in size (including a small pond). The entire area has been protected through the efforts of the Rhode Island Department of Environmental Management, The Nature Conservancy, and the Audubon Society of Rhode Island. This last organization sponsors a ranger to maintain surveillance on the area.

Rowdish Reservoir Floating Islands, adjacent to George Washington Memorial State Forest, near Chepachet. Several small peatland islands (totaling perhaps 20 acres) were created when the water level of a former pond was raised to make a large reservoir. Boggy embayments were pulled loose and floated to the center (much in the same way that the floating bog in Lake Sadawga, Vermont, was made; see the description there). The islands support black spruce, dwarf shrubs such as bog laurel and bog rosemary, and a variety of herbaceous plants.

Screech Hole Bog, near Slatersville. This long, 40-acre fen extends into Massachusetts. Drainage ways filtering through small areas of open water are bordered by sedge fen and a marginal red maple forest. An esker forms one boundary.

Widow Smith Road Bog, adjacent to Pascoag Reservoir, near Pascoag. This 3-acre quaking bog is of the classic kettlehole configuration but had its origin in an unglaciated depression. It is vegetatively zoned from open water to sedge-heath mat fringe, with a heavily shrubbed outer belt adjacent to the upland.

In addition to these peatlands, several large swamps are worthy of mention because they have peatland parts containing Atlantic white cedar, mountain laurel, rhododendron, highbush blueberry, sweet pepperbush, red maple, sedges, and others. Especially good examples are *Great Swamp* (Great Swamp Wildlife Reservation), near Kingston (3,200 acres); *Cedar Swamp*, near Bradford (1,000 acres); *Newton Swamp*, near Westerly (2,000 acres); and *Potts Bog*, near East Greenwich (400 acres).

Contacts and Further Information

Audubon Society of Rhode Island
40 Bowen Street
Providence, Rhode Island 02903

Rhode Island Department of
Environmental Management
Division of Planning and
Development
Natural Heritage Program
83 Park Street
Providence, Rhode Island 02903

MASSACHUSETTS

Atlantic White Cedar Swamp Trail (Marconi Site), Cape Cod National Seashore, near Wellfleet. Owned by National Park Service. An elevated, well-constructed boardwalk extends into and around a fine peatland/swamp on Cape Cod, dominated by stately Atlantic white cedar trees, mixed with some red maple. The understory has highbush blueberry, inkberry, swamp azalea, sheep laurel, and other shrubs. At the edges grow typical Cape upland species such as broom crowberry, pitch pine, black jack oak, and wintergreen. A descriptive brochure is available at the site.

Black Pond Bog (Vinal Nature Preserve), near Cohasset. Owned by The Nature Conservancy. This 50-acre preserve has a 25-acre quaking bog with a small central pond; it is ringed by a sedge, dwarf shrub, and sphagnum

Fig. 73 Ponkapoag Bog, Massachusetts.
(Photo by Charles W. Johnson.)

mat, all attractively enclosed by a youthful wooded peatland of Atlantic white cedar, winterberry, mountain holly, and several heaths.

Hawley Bog (Cranberry Swamp), near Hawley. A National Natural Landmark. Owned by the Five Colleges, Inc. (a consortium of Amherst College, Smith College, Mount Holyoke College, Hampshire College, and the University of Massachusetts). Permission is required for entry. This 100-acre peatland sits in a trapped drainage of the Berkshire Hills of western Massachusetts and has characteristic species for the region, including black spruce, leatherleaf, bog and sheep laurel, bog rosemary, carnivorous plants, and many others. It appears to have developed in an old glacial lake basin, since it has forty feet of peat in the middle.

Ponkapoag Bog, near Canton and Randolph. Owned by the Metropolitan District Commission, part of the Blue Hills Reservation. According to the sign at the beginning of the trail, this 50- to 60-acre peatland has "one of the first and longest log nature trails in the United States." The boardwalk, built in 1949 of native Atlantic white cedar, is still in good condition today. Nearest the edge the peatland is composed mostly of highbush blueberry, arrowwood, and Atlantic white cedar. After about a quarter of a mile it changes to an open bog with leatherleaf, sphagnum hummocks and hollows, cranberries, sundews, and stunted white cedar. The trail ends at a pond, where more mineral-rich waters support sweet gale, bladderworts, buckbean, and pitcher plants.

Thoreau's Bog (Gowing Swamp), near Concord. A classic kettlehole bog at least 36 feet deep with a quaking peat mat. Studied extensively, it contains species typical of a southern New England black spruce and tamarack peatland: sheep laurel, leatherleaf, bog rosemary, large and small cranberry, sundews, pitcher plants, and a variety of sphagnum mosses. The upland-wetland transition includes highbush blueberry, swamp azalea, sweet pepperbush, and huckleberry.

Charles W. Ward Reservation, near Andover. Owned by Trustees of Reservations. This 30-acre natural area has a boardwalk out onto a mat supporting black spruce, tamarack, sphagnum mosses, pitcher plants, and their normal associates for northeastern Massachusetts. The mat surrounds a tiny open pond. A booklet describing the area is available at the site.

Wolf Swamp, Cookson State Forest, near New Marlborough. Owned by the Massachusetts Department of Environmental Management. A 100-acre forested/shrub peatland, Wolf Swamp appears to have arisen as a shallow section of a larger pond (it now abuts East Indies Pond). It has a uniform overstory of black spruce and tamarack and an understory of Labrador tea, sheep laurel, leatherleaf, and some others. Most of the peatland floor consists of sphagnum.

Elsewhere in the state, excellent cedar swamps are *Hockamock Swamp* (Hockamock Swamp Wildlife Management Area), near Taunton (6,000 acres), owned by the Division of Fisheries and Wildlife; *Great Cedar Swamp*, near South Hanson (500 acres); and *Acushnet Cedar Swamp*, near New Bedford (1,000 acres), a National Natural Landmark.

The fall months (September to November) are a good time to see the cranberry bogs of southeastern Massachusetts (including Nantucket and Martha's Vineyard) as their crops are being harvested. The Cranberry World Visitor Center in Plymouth has comprehensive exhibits and information on cranberry processing, as well as directions to active bogs in the area.

Contacts and Further Information

Massachusetts Audubon Society
South Great Road
Lincoln, Massachusetts 01733

Massachusetts Department of
Environmental Management
100 Cambridge Street
Boston, Massachusetts 02202

Superintendent
National Park Service
Cape Cod National Seashore
South Wellfleet, Massachusetts
02663

NEW YORK

Bergen Swamp, near Bergen and Byron. A National Natural Landmark. Owned mostly by the Bergen Swamp Preservation Society, with a few private landowners. This 2,000-acre wetland complex has rich and diverse flora, varied landscapes, and interesting animals. Much of it is peatland with shallow, marly deposits. Northern white cedar grows at the wetland-upland interface and around seepages of calcium-rich waters. The calcareous waters also encourage growth of many unusual plants, including the rare white lady's slipper. In late summer when the water levels drop, the marl from the springs tends to cake up in sedge "marl meadows." Two boardwalks provide good entry.

Cicero Bog (Cicero Swamp), Cicero State Game Management Area, near North Syracuse. Owned by the New York Department of Conservation. Most of the 4,000-acre wetland is sphagnum-dominated peatland supporting black spruce, tamarack, stunted white pine, and a profuse spread of leatherleaf. A small, open, even more oligotrophic "island" of black huckleberry and Labrador tea, well removed from upland water sources, has developed in the middle of the forested peatland.

Conesus Lake Bog, Conesus Lake State Wildlife Management Area, near Conesus. Owned by the New York Department of Environmental Conservation. This 455-acre minerotrophic peatland has formed at the south end of Conesus Lake, likely formed in a long embayment of the original lake. Through a red maple swamp with thick grass and sedge mats multiple drainage ways lead to the lake, and shallow (less than ten feet deep) peats overlie a thick marl deposit on the old lakebed. The marly water indicates the minerotrophic conditions of the peatland. The whole area is circumscribed by roads, so access is easy.

Glen Lake Fen, near Lake George. This approximately 100-acre fen located behind the tourist-oriented Great Escape amusement park complex is well protected by a wide moat, which is actually a river oxbow. The vegetation is lush, with sweet gale, winterberry, bog willow, swamp rose, and numerous herbaceous plants. Of particular interest is the diversity of mosses, especially sphagna. Scattered tamaracks dot the interior portions of the peatland.

Heritage Road Bog, near Lake Bonaparte, Lewis County. This 900-acre wetland complex in roughly central New York contains several peatland

Fig. 74 Peatland near Glen Lake, New York. *(Photo by Charles W. Johnson.)*

types, most intriguing of which are the fens, with orchids, buckbean, tamarack, and other, even more uncommon fen species. Balsam fir grows nearby.

Junius Bog, Montezuma National Wildlife Refuge, near Mays Point and Montezuma. Owned by the U.S. Fish and Wildlife Service. Within a refuge of more than 6,000 acres, encompassing many wetlands—marshes, swamps, and peatlands—is a quaking bog of perhaps 25 to 35 acres. It is intermediate between bog and fen, with many plant species representative of both. The minerotrophic pond waters have thick marl deposits.

Kennedy Bog, Mendon Ponds Park, near Mendon. Owned by the Monroe County Parks Department. The peatland is part of a 1,000-acre wetland near Lake Ontario in the western part of the state and has both bog and fen characteristics and attendant flora. The peatland occurs in many excellent stagnant ice landforms created at the close of the Wisconsin Glaciation, notably kames, kettles, and eskers.

Lot Ten Bog, Lot Ten Swamp, near Palermo. Lot Ten Swamp is one of the numerous wetlands formed in northwestern New York's glaciated terrain in the Tug Hill Plateau and the lowlands bordering Lake Ontario. Ombrotrophic or oligotrophic bogs occupy kettles and blocked drainages between drumlins; fens and poor fens likewise are fairly common. Lot Ten Swamp is a 5,000-acre interdrumlin wetland; Lot Ten Bog is a 12-acre nutrient-poor bog within it, containing representative plants such as sphagnum mosses, leatherleaf, highbush blueberry, black chokeberry, tamarack, and black spruce.

McLean Bogs, Lloyd-Cornell McLean Wildlife Reservation, near Malloryville and McLean. Owned by Cornell University, Ithaca. Amid outstanding examples of recessional glacial landforms in the Finger Lakes region of New York—kames, eskers, and kettleholes—lie two bogs occupying a total of 81 acres. One is oligotrophic; the other, fed from springs, is minerotrophic. Minerotrophic Mud Pond Bog has fen species such as buckbean, marsh fern, skunk cabbage, and several orchids. The oligotrophic bog has more northern bog plants, especially heath shrubs and sedges. The upland around the bogs is mature northern hardwoods of good age and quality. This is an important teaching and research area for Cornell University.

Moss Lake Bog, near Houghton. Owned by The Nature Conservancy. This delightful peatland fulfills the classic northeastern imagery of a modestly large, basic-occupying bog complete with central pond and floating mat. Sphagnum mosses, heaths, bog clubmoss, pitcher plants, bladderworts, sundews, and orchids grow on the mat that surrounds the small lake.

Ringwood Bog, near Freeville. Owned by Cornell University. This small (approximately 20-acre) acidic bog is unusual for the Finger Lakes region; most of the peatlands here are minerotrophic, owing to the influence of the calcareous bedrock. Ringwood Bog has typical northern bog plants, including sphagnum mosses, leatherleaf, black spruce, and in more fenny areas, highbush blueberry.

Spring Pond Bog, Adirondack Preserve, near Saranac Lake. Under acquisition by The Nature Conservancy. Spring Pond Bog, covering 550 acres, is one of several large peatlands in the Adirondack Preserve. Most of the bog is ombrotrophic open peatland, parts of it dotted with scattered and widely spaced low trees. This is the only northeastern peatland outside of Maine containing both ridge and hollow patterning and surface ponds. *Sphagnum rubellum* dominates much of the mat surface, imparting a beautiful red to the expanse. Other plants include leatherleaf, bog laurel, sheep laurel, dwarf birch, black spruce, Labrador tea, and several sedges and orchids. Spatulate-leaved sundew borders some pond margins. Spring Pond Bog and neighboring wetlands provide habitat and refuge for many boreal birds, some such as spruce grouse, gray jays, black-backed woodpeckers, and common loons that are rare or absent in much of the Northeast. (This exceptionally significant bog is vulnerable to excessive foot traffic. Please contact the Adirondack Conservancy before visiting.)

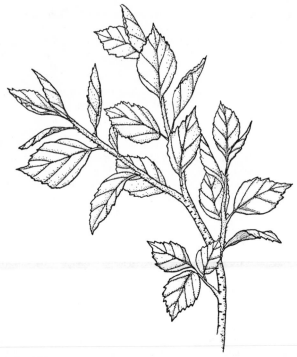

Fig. 75 Bog (dwarf) birch (*Betula pumila*). A peatland shrub rare throughout the Northeast (but widespread from Newfoundland to Michigan and south to New Jersey and northern Indiana), bog birch favors minerotrophic locales, such as calcareous fens and nutrient-rich laggs of bogs. Under optimum conditions it attains a height of 10 to 12 feet. Its 1-inch-long, egg-shaped leaves are often downy on both sides, especially when young.

Zurich Bog, near Zurich. Owned by the Bergen Swamp Preservation Society. This 490-acre peatland is long (2½ miles) and narrow (½ mile), lying between large drumlins. Encircling a drumlin, Zurich Bog borders a pond (Mud Pond) just to the north. The mat contains a variety of bog and fen flora, including water willow, highbush blueberry, mountain holly, black huckleberry, small cranberry, round-leaved sundew, and several species of orchids. Toward the upland the mat vegetation shifts to trees and shrubs indicative of more minerotrophic conditions: red maple, northern white cedar, spicebush, and others.

Contacts and Further Information

Adirondack Park Agency
P.O. Box 99
Ray Brook, New York 12977

The Nature Conservancy, New York chapters:

Adirondack Conservancy
P.O. Box 188
Elizabethtown, New York 12932

Eastern New York Chapter
1736 Western Avenue
Albany, New York 12203

Long Island Chapter
P.O. Box 72
Cold Spring Harbor, New York 11724

Lower Hudson Chapter
RFD 2
Mt. Kisco, New York 10549

South Fork Shelter Island Chapter
P.O. Box JJJJ
East Hampton, New York 11937

New York Department of Environmental Conservation
Wildlife Resources Center
Endangered Species Unit
Delmar, New York 12054

U.S. Fish and Wildlife Service
Montezuma National Wildlife Refuge
R.D. 1, Box 1411
Seneca Falls, New York 13148

VERMONT

Lake Carmi Bog, Lake Carmi State Park, near Enosburg Falls. Owned by the Vermont Department of Forests, Parks, and Recreation. This 230-acre forested bog formed in a bay of Lake Carmi. Its vegetative uniformity is unusual for the state: it is almost totally covered by tall and spindly black spruce, with lesser amounts of tamarack. Several shrubs form a thick and tough-to-penetrate understory; mountain holly is the most prevalent. The ground layer is not particularly rich but does have pitcher plants, moccasin flowers, three-leaved Solomon's seals, sedges, and species of sphagnum mosses.

Lake Sadawga Floating Bog, near Whitingham. Ownership uncertain. This anomaly—a 20-acre bog in the middle of a lake—apparently resulted from damming of the lake to raise the water level. A peatland that had formed in a bay was torn loose and drifted around the lake for several years. Now the island seems to be anchored by its peats. The vegetation at its outer edge is minerotrophic, with such species as sweet gale and arrow arum; inside this very unsteady zone is a firmer sphagnum mat supporting black spruce, tamarack, heaths, sedges, and carnivorous plants.

Lost Pond Bog, Green Mountain National Forest, by the Long Trail near Buckball Peak. Owned by the Green Mountain National Forest. This pristine 10- to 15-acre area is an example of a high-elevation (2,700 feet) pond

Fig. 76 Lake Carmi Bog, Vermont.
(Photo by Charles W. Johnson.)

with a bordering semifloating mat. Common plants here include sedges, sphagnum mosses, heaths, sundews, and pitcher plants.

Molly Bog, near Stowe. Owned by the University of Vermont. Permission to enter required. A National Natural Landmark. A classic northeastern kettlehole bog of perhaps 30 acres, with textbook zonation from open pond to evergreen forest. The small pond (about an acre) is ringed with a margin of sedge, leatherleaf, and sweet gale, which in turn is surrounded by a zone of sphagnum mosses, of which the red *S. rubellum* is outstanding. Pitcher plants, sundews, bladderworts, heaths, and sedges grow on the mat beyond the marginal zone. The bordering bog forest of black spruce and tamarack has a thick understory of highbush and black highbush blueberry, mountain holly, and rhodora.

Moose Bog, Wenlock Wildlife Management Area, near Island Pond. Owned by the Vermont Fish and Wildlife Department. This 30- to 40-acre peatland has a shallow open pond surrounded by a wide sphagnum-sedge mat, bordered by black spruce peatland forests an upland spruce and fir forests. Part of a 2,000-acre wildlife management area, it is home not only to an array of bog plants (including many heaths, pitcher plants, and sedge and sphagnum species) but also to uncommon boreal animals, especially birds. Nesting here are boreal chickadees, black-backed woodpeckers, spruce grouse, and gray jays. Moose are observed occasionally, and moose and bear sign are abundant.

Peacham Bog, Groton State Forest, near Marshfield. Owned by the Vermont Department of Forests, Parks, and Recreation. This acidic peatland of more than 200 acres is part of a 750-acre natural area preserve. Peacham Bog appears slightly raised (there are only two or three such in the state) and has a variety of northern plants, including rhodora, Labrador tea, small cranberry, leatherleaf, and bog rosemary, and sphagna growing on its hummocky mat. Isolated black spruce clusters show much layering. Tamarack and many sedge species occur in the more minerotrophic portions at the edge and in water tracks through the mat.

Snake Mountain Bog (Cranberry Bog), Snake Mountain Wildlife Management Area, near Weybridge. Owned by the Vermont Fish and Wildlife Department. A small (6- to 7-acre) basin bog sits atop the broad ridgetop of 1,287-foot Snake Mountain in the Lake Champlain Valley. An oligotrophic peatland, it contains many of the plants common in northern New England peatlands, including an exceptional density of pitcher plants.

Victory Bog, Victory Bog Wildlife Management Area, near Victory. Owned by the Vermont Fish and Wildlife Department. This 10- to 15-acre oligotrophic peatland is in the midst of a large and interesting complex of wetlands (sometimes collectively also known as Victory Bog) and uplands, all of which is a remote, wild hunting ground for many boreal and wildlands-loving animals, such as moose, coyote, otter, bobcat, and fisher. The peatland has widely spaced and stunted black spruce, with similarly dwarfed Labrador tea, bog laurel, and other heaths growing on a thick sphagnum mat. Access is difficult and unmarked.

Contacts and Further Information

Green Mountain National Forest
Court House Building
Rutland, Vermont 05701

The Nature Conservancy,
Vermont Chapter
138 Main Street
Montpelier, Vermont 05602

University of Vermont
Environmental Studies Program
The Bittersweet Building
Prospect and Main Streets
Burlington, Vermont 05405

Vermont Department of Forests,
Parks, and Recreation
79 River Street
Montpelier, Vermont 05602

Vermont Fish and Wildlife
Department
79 River Street
Montpelier, Vermont 05602

NEW HAMPSHIRE

Bradford Bog, near Bradford. Owned by the New England Wildflower Society and private owners. This 600-acre peatland in the southwest quarter of New Hampshire contains a stand of Atlantic white cedar and associated understory plants, including rhodora, bog rosemary, and leatherleaf. It has rich bird and mammal populations, as well.

Church Pond Bog, near Albany. Owned by the White Mountain National Forest, in the northern half of the state. This largely open bog has abundant leatherleaf, Labrador tea, bog laurel, small cranberry, and other associated species. A black spruce and tamarack forest occurs at the southern edge of the bog and farther back encircles the open mat and pond. Another small peatland is nearby. Many boreal birds such as boreal chickadees, spruce grouse, white-winged crossbills, and many warblers frequent the area.

Heath Pond Bog, Ossipee Lake, near Center Ossipee. Owned by the New Hampshire Division of Parks. This 100-plus-acre peatland in east-central New Hampshire contains a small pond near one side, surrounded by a narrow mat of sphagnum mosses, white beak-rush, leatherleaf, small cranberry, sheep laurel, and rhodora, and liberal swaths of Labrador tea and bog laurel throughout. Fen species, such as highbush blueberry and maleberry, and carnivorous plants and orchids, are all present. Black spruce and tamarack are scattered over the bog. The rich flora, the presence of both northern and southern plant species, and the unspoiled quality are why Heath Pond Bog is a National Natural Landmark.

Moose Pasture Bog, near Pittsburg. Owned by the St. Regis Paper Company. This 100-acre peatland in extreme northern New Hampshire supports a black spruce and tamarack forest with a thick understory of shrubs such as mountain holly, black chokeberry, and Bartram's shadbush. Many heaths, sedges, carnivorous species, and orchids also grow here. The site is equally interesting for the great variety of boreal birds that visit or nest: spruce grouse, boreal chickadees, crossbills, gray jays, and several species of uncommon warblers, among others. The peatland is part of the 336-acre East Inlet Natural Area, which includes a large old-growth stand of red spruce.

Mud Pond Bog, Fox State Forest, near Hillsborough. Owned by the New Hampshire Division of Forests and Lands. This 50-acre poor fen within the state forest has clear zonation, with open water surrounded by a wide mat. Where the mat extends into the water, water willow, arrow arum, and small "islands" of bog rosemary grow. Farther back and away from the water, conditions are slightly more acidic, with small cranberry, sundews, bog laurel, bog rosemary, leatherleaf, sheep laurel, and a variety of sphagnum mosses. The periphery of the mat has highbush blueberry, buttonbush,

Fig. 77 Mud Pond Bog, New Hampshire. *(Photo by Charles W. Johnson.)*

black chokeberry, and others growing thickly. Virginia chain fern is also found there.

Pondicherry Wildlife Refuge, near Jefferson and Whitefield. Owned jointly by the New Hampshire Fish and Game Department and the Audubon Society of New Hampshire, the refuge includes Cherry Pond Bog (235 acres) and Little Cherry Pond Bog (69 acres). Both peatlands surround open ponds and have some open mat of sphagnum mosses, accompanying heaths, and other low vegetation. Black spruce and tamarack adjoin the mat. The refuge is rich with bird life, including many boreal and marsh/swamp species.

Ponemah Bog, near Nashua. Owned by the Audubon Society of New Hampshire. About 100 acres of poor fen surround a 3-acre pond. The mat supports a profusion of rhodora and bog rosemary. Some peat has been cut out of the bog in the past for domestic use. It also has a forest of red maple at the upland-wetland interface. The Audubon Society has constructed a boardwalk through the area for easy access and observation. Ponemah Bog is in extreme southern New Hampshire, not far from the Massachusetts border.

Smith Pond Bog, near Hopkinton. Owned by the Audubon Society of New Hampshire and the New Hampshire Fish and Game Department. 55 acres. A large pond is surrounded by a narrow open mat, which quickly merges with a wider, shrubby zone. The entire peatland (in central New Hampshire) is poor fen, with such species as sweet gale, black alder, water willow, and highbush blueberry abundant. Both large and small cranberry are here, as well as sundew and pockets of pitcher plants. Beavers have been active in the pond and may have affected the character of the peatland. Access is from several points; one has a boardwalk.

Contacts and Further Information

Audubon Society of New Hampshire
3 Silk Farm Road
Concord, New Hampshire 03301

New England Wildflower Society
Hemenway Road
Framingham, Massachusetts 01701

New Hampshire Department of Fish and Game
Concord, New Hampshire 03301

New Hampshire Office of State Planning
2½ Beacon Street
Concord, New Hampshire 03301

St. Regis Paper Company
West Stewartstown, New Hampshire 03597

Society for the Protection of New Hampshire Forests
54 Portsmouth Street
Concord, New Hampshire 03301

MAINE

Appleton Bog, near Appleton. A National Natural Landmark. Owned by The Nature Conservancy and other landowners. This 630-acre site includes a 230-acre stand of virgin or near-virgin Atlantic white cedar, unusual elsewhere on the Atlantic coast. Except for a meager colony near Rockport, the site marks the northernmost limit of the species in North America. Numerous other wetland types also occur here: a red maple swamp, a 32-acre pond, and a quaking bog with sundews, pitcher plants, heaths, Virginia chain fern, black spruce, and tamarack.

Big Heath, Acadia National Park, Mt. Desert Island, near Southwest Harbor. Owned by the National Park Service. This 420-acre plateau peatland, moderate in size for its type, is the southernmost example in North America. The central part of the bog is open, with black crowberry, sphagnum mosses (particularly *S. fuscum*), deer's-hair sedge, and baked-apple berry—all characteristic of the plateaus. Black spruce dots the surface in clusters and surrounds most of the perimeter. The numerous ponds and wet depressions in the open peatland are the southern limit of these features, so prominent at more northerly latitudes. A short trip by boat from Southwest Harbor, the 215 acres of *Great Cranberry Isle Heath* on Great Cranberry Isle have much the same flora as the Big Heath, but the pleateau structure is much clearer and more obvious.

Caribou Bog (Orono Bog), near Orono and Old Town. This complex 2,500-acre peatland has been subjected to many different uses over the years: hunting, fishing, nature study, route for a railroad and a pipeline, town dumps, and research/education by the nearby University of Maine personnel. Most of the peatland is a mildly raised bog, with slight doming in one section, dotted with a few secondary pools. It also has some minerotrophic sections. Much of the surface is forested and/or shrubbed.

Carrying Place Cove, near Lubec. This 43-acre peatland in extreme northeastern coastal Maine, small by other plateau standards, is a very special place. The flora of the bog is similar to that of other pleateaus, but the main attraction is at the shore, where the ocean waves are eating into the bog and exposing the entire 10- to 15-foot-deep peat profiles and underlying mineral substrate. Because of its well-developed and visible features, Worley (1981) rated it "one of the most significant peatlands in the eastern United States." Its value is acknowledged with the designation as a National Natural Landmark.

Crystal Bog, near Crystal and Sherman. A National Natural Landmark. Owned mostly by The Nature Conservancy. This 1,550-acre northern Maine peatland is complex, containing many features: a domed bog section with the finest array of concentrically patterned surface pools in the United States outside of Alaska, unpatterned raised bog portions, and several open-basin minerotrophic fen bog areas. One of the most fascinating and significant peatlands in eastern North America, Crystal Bog has a rich and varied flora, especially in the rich fen sections, which contain some of the most uncommon of the Northeast's peatland plants. Because it is such a significant yet sensitive place, visitors should take special care not to disturb the peatland.

Great Heath, near Columbia and Harrington. Owned largely by the Maine Department of Conservation, with about one-third in private hands. At 3,430 acres, the central expanse is Maine's largest continuous open bog. An additional 911 acres of peripheral peatland gives the area of peatland known as Great Heath a total of more than 4,300 acres. Its surface is a complex coalesced dome, strongly patterned in places, with pool systems and soaks, and it has an immensity unrivaled by any other northeastern bog. In its structure and vegetation, the Great Heath appears to be intermediate between inland and maritime raised bogs. Like inland bogs it is domed, has small islands of trees and extensive dwarf shrub communities, and is dotted with pools. But like maritime raised bogs it has some areas with lawns of deer's-hair sedge and the black crowberry-dwarf huckleberry community (which also contains baked-apple berry, considered a rare species in Maine). Initial surveys list more than 75 wetland vascular plant species. Immediately south of the Great Heath is one of the best preserved, emerged glaciomarine deltas in the United States. In the last few years the Great Heath has been the subject of several resource and scientific investigations, yielding some fine descriptions and baseline information. Its

Fig. 78 Crystal Bog, Maine. *(Photo courtesy of Maine Critical Areas Program.)*

character, size, and pristineness make it one of the most significant peatlands of the Northeast.

Great Wass Peatlands, Great Wass Island, near Jonesport. Owned by The Nature Conservancy. Because it juts so abruptly out from the general coastline into the Gulf of Maine, and since it lies so far to the northeast in Maine, Great Wass Island has an extremely foggy, humid, cool maritime climate. Scattered over its lumpy topography are 180 or so acres of peatlands. Sharply different from place to place on the island, the peatlands range from ombrotrophic to minerotrophic. One or two raised bogs form rather indistinct plateaus with abundant black crowberry, baked-apple berry, and deer's-hair sedge. Elsewhere, wet graminoid fens prevail. Jack pine grows directly on the peatlands; it does so nowhere else in the Northeast. Over the rounded bedrock of the most seaward tip of the island are blanketing slope peatlands. Covered with dwarf shrubs and occasional sphagna and lichens, the peats, though seldom saturated, are kept almost perpetually moist by fog, mists, or rain.

Jonesport Heath, near Jonesport. This peatland complex of about 1,000 acres contains the largest coastal plateaus in the United States. At this writing, mining operations (for horticultural peat) have spread over most of the peatland; however, the northernmost plateau remains essentially undisturbed. The plateaus are treeless with central lawns of deer's-hair sedge and *Sphagnum fuscum*, bordered by much black crowberry and baked-apple berry. The marginal slopes have low shrubs of sheep laurel, rhodora, and sweet gale. The zonation of the communities and the plateaus are striking.

Mahoosuc Mountain Bogs, Mahoosuc Range. In the extreme western part of Maine in Oxford County, next to the New Hampshire border, is the Mahoosuc Range, part of the White Mountain system and traversed by the Appalachian Trail. The Maine Department of Conservation owns a 21,000-acre stretch of the Mahoosucs, including some rugged peaks around 4,000 feet high and several small (less than an acre or so) subalpine peatlands. These wet sedge- and moss-dominated peatlands fill depressions along the high ridges, grading with the soils and vegetation of another treeless community—the moist subalpine heathlands (sometimes called "heath balds"). With the sedges and sphagnum (the main peat forms) there commonly are hare's-tail cotton grass, lichens, baked-apple berry (a plant rare in the United States outside of Alaska, elsewhere found on the coastal bogs of eastern Maine), deer's-hair sedge, and occasional sundews. In the wettest areas is small cranberry. Boardwalk trails through some of these bogs protect the vegetation while giving visitors a dry passage and close looks.

Meddybemps Heath, near Calais. A National Natural Landmark. A large coalesced domed bog complex of some 2,000 acres, Meddybemps Heath is quite similar in morphology, patterning, and flora to the Great Heath. Several areas of bog pools occur on the domes amid hummocky communities of stunted black huckleberry, leatherleaf, sheep laurel, and other heaths. Sphagnum is a principal peat former; *S. fuscum* is a major hummock species. The many water tracts, soaks, and streams on or by the bog have diverse vegetation; in shrubby areas there are often rhodora, leatherleaf, mountain holly, or alder. On apparently drier areas of the domes black crowberry may be prominent. A fine esker forms part of the eastern border of the bog. Meddybemps Heath lies along the western shore of Meddybemps Lake, which provides a scenic water access to the bog.

Number 5 Bog, near Jackman, Somerset County. A National Natural Landmark. Perhaps the most remote of the Northeast's largest peatlands, Number 5 Bog is pristine, essentially untouched by man, and until very recently virtually unknown. Nestled beneath the mountains of western Maine, bordered by glacial moraines, near to lakes and rivers, Number 5 Bog holds outstanding aesthetic appeal. With some 1,500 acres, the 10,500-year-old peatland consists of two adjacent areas, the largest of which has a substantial (90 acres) central pond. Unlike most other large Maine peatlands, much of Number 5 Bog is very wet, with many sedges, zones of widely spaced tamarack, and wet sphagnum lawns with much *S. rubellum.* Its aligned ponds and ridges in parallel formation (some parallel to the edge of the central pond) indicate surface water flow and are a type of ribbed fen. No other peatland in the Northeast has such a combination of features. Wooded areas by soaks, streams, and around the peatland have northern white cedar with moss-rich understories. The immense areas of wet and somewhat nutrient-enriched peatland provide excellent habitat for several colorful orchids (e.g., arethusa and white-fringed orchid), a variety of sedges (many *Carex* species), and many broad-leaved herbs such as buckbean and three-leaved Solomon's seal. The nutrient-poor bog communities have common heaths (e.g., leatherleaf, Labrador tea, sheep laurel) and hummocks of sphagnum. A preliminary survey lists 57 species of vascular plants from the wetlands. A significant population of jack pine, along with red pine, grows in the uplands around the bog. Regarded as one of the most important peatlands in the Northeast, this bog has been protected by its isolation and lack of easy access. A Maine Critical Area, it now is afforded some extra protection through ownership by a private conservation organization.

Saco Heath, near North Saco, in southernmost Maine. This peatland of about 500 acres has some bog features indicative that portions are slightly raised, marginal, somewhat enriched fens with a faint hint of rib patterning, and a stand of Atlantic white cedar. Many of Maine's peatland plants grow at Saco Heath—from black spruce, tamarack, and heaths to sedges, carnivores, and a number of sphagna. Little studied, the site may be the southernmost for some of its morphological characters. A small part has been mined for peat, and some cedars have been cut. A larger mining operation is being considered.

Contacts and Further Information

Maine Audubon Society
Gilsland Farm
118 Route 1
Falmouth, Maine 04105

Maine Critical Areas Program
State Planning Office
184 State Street
Augusta, Maine 04333

Maine Department of
Conservation
State House Station 222
Augusta, Maine 04333
(This department includes the
Bureau of Forestry, Maine
Geological Survey, and Bureau of
Public Lands.)

Maine Department of Inland
Fisheries and Wildlife
284 State Street, Station 41
Augusta, Maine 04333

The Nature Conservancy, Maine
Chapter
20 Federal Street
Brunswick, Maine 04011

APPENDIX 2 | SOME AMPHIBIANS AND REPTILES OF NORTHEASTERN PEATLANDS

The following species, listed in phylogenetic order but not including subspecies, are found in the general region. The list is based on Stockwell and Hunter (1983), Minnesota Department of Natural Resources (1981), and New Jersey Pinelands Commission (1980). These species may be found in other habitats, sometimes commonly.

	Open Bog	Open Fen	Shrub Thicket	Forested Peatland
Jefferson's salamander (*Ambystoma jeffersonianum*)			x	
Blue-spotted salamander (*Ambystoma laterale*)	x	x	x	x
Eastern (red-spotted) newt (*Notophthalmus viridescens*)	x	x		
Four-toed salamander (*Hemidactylium scutatum*)	x	x	x	x
Spring salamander (*Gyrinophilus porphyriticus*)				x
Mud salamander (*Pseudotriton montanus*)			x	x
Red salamander (*Pseudotriton ruber*)			x	x
Two-lined salamander (*Eurycea bislineata*)				x

	Open Bog	Open Fen	Shrub Thicket	Forested Peatland
American toad (*Bufo americanus*)	x	x	x	x
Northern cricket frog (*Acris crepitans*)		x		
Spring peeper (*Hyla crucifer*)		x	x	x
Pine Barrens tree frog (*Hyla andersoni*)	x	x	x	x
Gray tree frog (*Hyla versicolor*)			x	x
Bullfrog (*Rana catesbeiana*)			x	x
Carpenter frog (*Rana virgatipes*)	x	x		
Green frog (*Rana clamitans*)		x	x	
Mink frog (*Rana septentrionalis*)	x	x		
Wood frog (*Rana sylvatica*)	x	x	x	x
Northern leopard frog (*Rana pipiens*)		x	x	x
Pickerel frog (*Rana palustris*)		x	x	
Snapping turtle (*Chelydra serpentina*)		x	x	
Stinkpot (*Sternotherus odoratus*)				x
Spotted turtle (*Clemmys guttata*)		x		
Bog turtle (*Clemmys muhlenbergii*)	x	x	x	
Painted turtle (*Chrysemys picta*)	x	x		
Blanding's turtle (*Emydoidea blandingii*)	x	x		
Northern water snake (*Nerodia sipedon*)	x	x	x	x
Redbelly snake (*Storeria occipitomaculata*)	x			x
Common garter snake (*Thamnophis sirtalis*)	x	x	x	x
Eastern ribbon snake (*Thamnophis sauritus*)	x	x	x	x

	Open Bog	Open Fen	Shrub Thicket	Forested Peatland
Rough green snake (*Opheodrys aestivus*)	x		x	x
Common kingsnake (*Lampropeltis getulus*)				x
Milk snake (*Lampropeltis triangulum*)				x
Massasauga (*Sistrurus catenatus*)		x		x

APPENDIX 3 | SOME BIRDS OF NORTHEASTERN PEATLANDS

The following species, listed in phylogenetic order, are found in the general region during the breeding season. The list is based on Stockwell and Hunter (1983), Minnesota Department of Natural Resources (1981), and New Jersey Pinelands Commission (1980). These species may be found in other habitats, often commonly.

	Open Bog	Open Fen	Shrub Thicket	Forested Peatland
Common loon (*Gavia immer*)		x		
Pied-billed grebe (*Podilymbus podiceps*)		x	x	
American bittern (*Botaurus lentiginosus*)		x		
Great blue heron (*Ardea herodias*)		x	x	x
Snowy egret (*Egretta thula*)		x	x	
Green-backed heron (*Butorides striatus*)			x	x
Black-crowned night heron (*Nycticorax nycticorax*)			x	x
Glossy ibis (*Plegadis falcinellus*)		x	x	
Wood duck (*Aix sponsa*)			x	x

	Open Bog	Open Fen	Shrub Thicket	Forested Peatland
American black duck (*Anas rubripes*)		x	x	
Mallard (*Anas platyrhynchos*)		x	x	
Blue-winged teal (*Anas discors*)		x		
Hooded merganser (*Lophodytes cucullatus*)			x	x
Common merganser (*Mergus merganser*)		x	x	
Osprey (*Pandion haliaetus*)		x		x
Bald eagle (*Haliaeetus leucocephalus*)		x		x
Northern harrier (*Circus cyancus*)	x	x		x
Cooper's hawk (*Accipiter cooperii*)			x	x
Red-shouldered hawk (*Buteo lineatus*)			x	x
Broad-winged hawk (*Buteo platypterus*)			x	x
American kestrel (*Falco sparverius*)	x	x		
Merlin (*Falco columbarius*)				x
Peregrine falcon (*Falco peregrinus*)		x		
Spruce grouse (*Dendragapus canadensis*)				x
Ruffed grouse (*Bonasa umbellus*)			x	x
Yellow rail (*Coturnicops noveboracensis*)		x		
Sora (*Porzana carolina*)		x		
Killdeer (*Charadrius vociferus*)		x		
Spotted sandpiper (*Actitis macularia*)		x		
Common snipe (*Gallinago gallinago*)	x	x	x	x
American woodcock (*Scolopax minor*)		x	x	x

	Open Bog	Open Fen	Shrub Thicket	Forested Peatland
Ring-billed gull (*Larus delawarensis*)	x	x		
Herring gull (*Larus argentatus*)	x	x		
Mourning dove (*Zenaida macroura*)		x	x	x
Black-billed cuckoo (*Coccyzus erythropthalmus*)			x	x
Barred owl (*Strix varia*)				x
Ruby-throated hummingbird (*Archilochus colubris*)	x	x	x	
Belted kingfisher (*Ceryle alcyon*)		x		x
Yellow-bellied sapsucker (*Sphyrapicus varius*)				x
Downy woodpecker (*Picoides pubescens*)				x
Hairy woodpecker (*Picoides villosus*)				x
Black-backed woodpecker (*Picoides arcticus*)				x
Northern flicker (*Colaptes auratus*)			x	x
Pileated woodpecker (*Dryocopus pileatus*)				x
Olive-sided flycatcher (*Contopus borealis*)			x	x
Eastern wood-pewee (*Contopus virens*)				x
Yellow-bellied flycatcher (*Empidonax flaviventris*)				x
Alder flycatcher (*Empidonax alnorum*)			x	
Least flycatcher (*Empidonax minimus*)			x	
Eastern phoebe (*Sayornis phoebe*)			x	x
Great crested flycatcher (*Myiarchus crinitus*)			x	x
Eastern kingbird (*Tyrannus tyrannus*)	x	x		
Tree swallow (*Tachycineta bicolor*)	x	x	x	

	Open Bog	Open Fen	Shrub Thicket	Forested Peatland
Bank swallow (*Riparia riparia*)	x	x		
Cliff swallow (*Hirundo pyrrhonota*)		x		
Barn swallow (*Hirundo rustica*)	x	x		
Gray jay (*Perisoreus canadensis*)	x	x	x	
Blue jay (*Cyanocitta cristata*)			x	x
Common raven (*Corvus corax*)				x
Black-capped chickadee (*Parus atricapillus*)			x	x
Boreal chickadee (*Parus hudsonicus*)				x
Red-breasted nuthatch (*Sitta canadensis*)				x
Brown creeper (*Certhia americana*)				x
House wren (*Troglodytes aedon*)			x	
Winter wren (*Troglodytes troglodytes*)				x
Sedge wren (*Cistothorus platensis*)	x	x	x	
Golden-crowned kinglet (*Regulus satrapa*)				x
Ruby-crowned kinglet (*Regulus calendula*)				x
Eastern bluebird (*Sialia sialis*)	x	x		
Veery (*Catharus fuscescens*)			x	x
Swainson's thrush (*Catharus ustulatus*)				x
Hermit thrush (*Catharus guttatus*)				x
American robin (*Turdus migratorius*)			x	
Gray catbird (*Dumetella carolinensis*)			x	
Cedar waxwing (*Bombycilla cedrorum*)			x	x

	Open Bog	Open Fen	Shrub Thicket	Forested Peatland
Solitary vireo (*Vireo solitarius*)				x
Warbling vireo (*Vireo gilvus*)			x	x
Red-eyed vireo (*Vireo olivaceus*)			x	x
Golden-winged warbler (*Vermivora chrysoptera*)			x	
Tennessee warbler (*Vermivora peregrina*)			x	x
Nashville warbler (*Vermivora ruficapilla*)			x	x
Northern parula (*Parula americana*).				x
Yellow warbler (*Dendroica petechia*)			x	
Chestnut-sided warbler (*Dendroica pensylvanica*)			x	
Magnolia warbler (*Dendroica magnolia*)				x
Cape May warbler (*Dendroica tigrina*)				x
Yellow-rumped warbler (*Dendroica coronata*)	x			x
Black-throated green warbler (*Dendroica virens*)				x
Blackburnian warbler (*Dendroica fusca*)				x
Palm warbler (*Dendroica palmarum*)	x			x
Black-and-white warbler (*Mniotilta varia*)			x	x
American redstart (*Setophaga ruticilla*)			x	x
Ovenbird (*Seiurus aurocapillus*)				x
Northern waterthrush (*Seiurus noveboracensis*)			x	x
Mourning warbler (*Oporornis philadelphia*)				x
Common yellowthroat (*Geothlypis trichas*)		x	x	x
Wilson's warbler (*Wilsonia pusilla*)			x	

	Open Bog	Open Fen	Shrub Thicket	Forested Peatland
Canada warbler (*Wilsonia canadensis*)			x	x
Rose-breasted grosbeak (*Pheucticus ludovicianus*)			x	
Savannah sparrow (*Passerculus sandwichensis*)	x	x		x
Henslow's sparrow (*Ammodramus henslowii*)	x			
Song sparrow (*Melospiza melodia*)			x	
Lincoln's sparrow (*Melospiza lincolnii*)		x	x	x
Swamp sparrow (*Melospiza georgiana*)		x	x	
White-throated sparrow (*Zonotrichia albicollis*)			x	
Dark-eyed junco (*Junco hyemalis*)	x			
Bobolink (*Dolichonyx oryzivorus*)	x	x		
Red-winged blackbird (*Agelaius phoeniceus*)		x		
Rusty blackbird (*Euphagus carolinus*)			x	x
Common grackle (*Quiscalus quiscula*)			x	
Brown-headed cowbird (*Molothrus ater*)			x	x
Purple finch (*Carpodacus purpureus*)				x
White-winged crossbill (*Loxia leucoptera*)			x	x
Pine siskin (*Carduelis pinus*)			x	
American goldfinch (*Carduelis tristis*)	x			x
Evening grosbeak (*Coccothraustes vespertinus*)			x	x

APPENDIX 4 | SOME MAMMALS OF NORTHEASTERN PEATLANDS

The following species, listed in phylogenetic order, are found in the general region. The list is based on Stockwell and Hunter (1983), Minnesota Department of Natural Resources (1981), and New Jersey Pinelands Commission (1980). These species may be found in other habitats, sometimes commonly.

	Open Bog	Open Fen	Shrub Thicket	Forested Peatland
Masked shrew (*Sorex cinereus*)	x	x	x	x
Water shrew (*Sorex palustris*)	x	x		
Smoky shrew (*Sorex fumeus*)				x
Pygmy shrew (*Sorex hoyi*)	x	x	x	x
Short-tailed shrew (*Blarina brevicauda*)	x	x	x	x
Least shrew (*Cryptotis parva*)		x		
Eastern mole (*Scalopus aquaticus*)		x	x	
Star-nosed mole (*Condylura cristata*)			x	x
Snowshoe hare (*Lepus americanus*)			x	x

	Open Bog	Open Fen	Shrub Thicket	Forested Peatland
Eastern chipmunk (*Tamias striatus*)				x
Red squirrel (*Tamiasciurus hudsonicus*)				x
Southern flying squirrel (*Glaucomys volans*)				x
Northern flying squirrel (*Glaucomys sabrinus*)				x
Beaver (*Castor canadensis*)			x	x
Deer mouse (*Peromyscus maniculatus*)	x			x
White-footed mouse (*Peromyscus leucopus*)		x	x	
Meadow vole (*Microtus pennsylvanicus*)	x	x	x	x
Southern red-backed vole (*Clethrionomys gapperi*)	x	x	x	x
Muskrat (*Ondatra zibethicus*)		x	x	
Southern bog lemming (*Synaptomys cooperi*)	x			x
Northern bog lemming (*Synaptomys borealis*)	x	x		
Meadow jumping mouse (*Zapus hudsonius*)	x	x	x	
Porcupine (*Erethizon dorsatum*)				x
Coyote (*Canis latrans*)			x	x
Red fox (*Vulpes vulpes*)		x	x	
Black bear (*Ursus americanus*)				x
Raccoon (*Procyon lotor*)		x	x	x
Marten (*Martes americana*)				x
Fisher (*Martes pennanti*)			x	x
Ermine (*Mustela erminea*)	x	x	x	
Long-tailed weasel (*Mustela frenata*)	x	x	x	x

	Open Bog	Open Fen	Shrub Thicket	Forested Peatland
Mink (*Mustela vison*)		x	x	
Lynx (*Felis lynx*)			x	x
Bobcat (*Felis rufus*)			x	x
White-tailed deer (*Odocoileus virginianus*)			x	x
Moose (*Alces alces*)			x	x

INDEX

For the sake of brevity, Latin names have not been included, and plant and animal names appearing in Appendix 1 have not been listed. However, Latin names may be found in the text in conjunction with their common names.